The Country Canal

The Country Canal
Ronald Russell

DAVID & CHARLES
Newton Abbot London

Frontispiece
The Lock. A wood engraving by the Dalziell brothers after Miles Birket Foster

Title page
The northern branch of the Bude Canal at Putford in the 1960s

British Library Cataloguing in Publication Data

Russell, Ronald, 1924–
    The country canal
    1. Great Britain. Canals, history
    I. Title
    386.460941

ISBN 0–7153–9169–0

Typeset by Typesetters (Birmingham) Ltd
and printed in Hong Kong
by Wing King Tong Co Ltd
for David & Charles plc
Brunel House Newton Abbot Devon

Contents

	Introduction	7
1	Canals and Country Life	11
2	Canalscape	27
3	Life on Board	47
4	The Transformation of a Canal Village	57
5	The Canal Village Through the Years	68
6	A Fenland Voyage	86
7	Lost Country Canals	108
8	Over the Hills and Far Away	129
	Bibliography	156
	Acknowledgements	158
	Index	159

The first boat arrives (see page 66)

Introduction

The country canal – a ribbon of water winding through green fields, a grassy towing-path and rampant hedge alongside, trees here and there, a river close by, hills rising in the distance, thatched cottages, a church tower . . . now comes a lock with old wooden beams, worn stone steps, iron paddle gear, leaky gates, a cottage alongside, a smiling lock-keeper. . . A smiling lock-keeper? Lock-keepers do smile, of course – but what is one doing here at this single lock where maybe only a dozen boats pass during the day? Present or past? Fact or fiction? And where are we – or where might we be? The Oxford Canal? The Grantham Canal sixty years ago? The Herefordshire & Gloucestershire Canal in the middle of the last century? Or have we stepped into a Victorian painting by Birket Foster or Frederick William Watts, romantic, nostalgic, unbelievably pretty?

For a moment leave it there. Here comes a boat, towed by a bay horse, his brasses sparkling in the sun, the coloured wooden bobbles on his harness swinging against his sides. A small boy leads him, scruffily dressed but fairly clean, big-booted, chewing a straw. The boatman leans on the tiller; he wears a plush waistcoat, striped shirt, moleskin trousers and smokes a stubby pipe. His wife, bonnetted and aproned, stands in the cabin doorway. There are paintings of Gothic castles on the cabin sides and two water cans decorated with roses, daisies and geometrical patterns stand on the cabin roof. In the hold are baulks of timber – they look like railway sleepers . . .

And so they are. And in two or three years' time the railway will be complete and open, running nearly parallel to the canal through the same beautiful stretch of countryside, connecting the same market towns and villages that depended on the canal for trading and transport. The lock-keeper will move to the town, perhaps to become a ticket-collector or a porter; his cottage will become derelict and in time will be demolished. The boatman and his wife will struggle on for a few more years; then they will sell their boat and try to find another occupation. Their son will go to work in a factory; their horse will plod the roads for a few more years and then be taken to the knacker's.

And on my shelves is David St John Thomas's handsomely

(Above) Horse-drawn boats and engineer's launch, near Braunston, 1910 (Water-ways Museum)

(Opposite) Cruising on the Brecknock & Abergavenny Canal (British Waterways Board)

presented and widely read book *The Country Railway*, mourning the passing of so many branch lines – but with only one sentence about the canals which many of those lines forced into disuse and decay.

However, the whirligig of time has his revenges. Take, for example, the Caldon Canal which is a branch of the Trent & Mersey Canal, at one time with an extension to Uttoxeter, and which for many years had a flourishing trade in limestone. Several miles of the Caldon run through the spectacular Churnet Valley, claimed to be one of the most scenically beautiful lengths of waterway in Britain. But in 1847 the Uttoxeter extension was closed and a railway was laid

over much of its bed – though this too disappeared many years ago. And in the early twentieth century another line was opened through the valley close to the canal, taking almost all of its trade away, so by mid-century the Caldon was disused and in danger of closure. But local opinion would not allow this, and after years of debate and discussion a full restoration programme was begun in 1972; the canal was reopened two years later. Now there are many pleasure cruisers based on the Caldon, a horse-drawn boat makes passenger trips from Froghall, there are walkers on the towing-path and hire boats in the summer. The railway alongside, however, is deserted, except for a little freight train which once a day chugs along to the copper works at Froghall.

The country canal, then, lives on, sometimes as part of the interconnecting network of waterways, sometimes isolated but very active, like the Brecknock & Abergavenny. Some, such as the Grantham Canal, have water but no boats; others have neither but may still be followed and enjoyed – the Leominster Canal is one. Lastly, there are those such as the Kennet & Avon and the Montgomeryshire canals which are being brought back to full and vigorous life as channels of pleasure and recreation.

That so many country canals are still here to be explored and enjoyed is a quirk of history for which we must be grateful. These quiet corridors through our landscape – and we can find them from Cornwall to Aberdeen, from Hythe to Fort William – are precious, not simply as memorials of the past but as joys, we hope, for ever.

On the Croydon Canal

1 Canals and Country Life

It is easy to exaggerate the influence of canals on country life. In his study *Rural Life in Victorian England* Professor G. E. Mingay points out that they had relatively little impact:

> The farmers made use of them, of course, for sending produce to market, as did manufacturers of bricks and pottery, and indeed some farmers had helped to promote them, but in general the barges which moved slowly through the countryside were not part of it. Their primary cargo was coal, and their destinations the rising centres of industry.

Sometimes new settlements developed by waterway junctions, or existing settlements were greatly enlarged, but usually one or two wharves would be constructed at points convenient to a nearby village, with a wharfinger's house and yard and possibly a coal store. As trade increased so more wharves would be constructed, some owned by the canal company and available for public use and others privately built and operated. Now almost all of these have long disappeared, though the sites of several of them are recorded on waterway maps and in L. A. Edwards' gazetteer *Inland Waterways of Great Britain*.

Nevertheless several canals were promoted largely to serve the needs of the local farms and small country markets, deriving their profits therefrom. For example, one of the main cargoes on the Bude Canal was shelly sand for use as fertiliser, while the Wilts & Berks carried corn and cheese to city markets. As it happened, canals which relied mainly on a country trade were in general financially unsuccessful, but for their country customers they served a valuable purpose.

At its simplest the country wharf might be just a few planks laid by the water's edge and used by a single farmer, who might himself own a boat suitable for short-haul journeys. Larger wharves, especially if they were to handle coal, timber or building materials, would be more solidly constructed with storage areas and a small warehouse. Here the canal company might employ a wharfinger both for security and for the collection of charges and tolls; a road link would be needed and a bridge might have to be built. In a few places the wharf became the focus for a small settlement – for example Stoke Prior on the Worcester & Birmingham Canal which has lock, wharf, warehouse and a row of attractive cottages facing the canal.

It is not easy to establish exactly how many wharves there might have been on any particular canal; small private wharves especially came and went as needs fluctuated. The Bude Canal had nine principal wharves, but clearly in its 35½ miles (56km) length there would have been many more private offloading points for sand for the farmers' fields. The Wey & Arun Junction had only six main wharves in its 18½ miles (29km) but much of its line ran through wooded country.

The three southern east–west routes make an interesting comparison. The Kennet & Avon Canal, 57 miles (91km) long, owned nineteen wharves and also leased thirteen to private owners. There were thirty-six wharves on the winding Wilts & Berks Canal which had 51 miles (82km) of main line and nearly 16 miles (25km) of branches. The Thames & Severn however, with 30 miles (48km) of waterway owned only twelve wharves although the large inland port at Brimscombe compensated for this to some extent – here there were three wharves, several warehouses and offices, a boat-weighing machine, a large and handsome house for the agent with various reception rooms, twenty-six bedrooms and stabling, and a basin about 700ft (213m) in length. At Brimscombe goods could be transhipped from Severn trows to Thames barges and vice versa. The canal company provided substantial houses for its agents at other points on its line, too – as at Kempsford, Cricklade, Latton and Cirencester, and it added another at Lechlade when it moved its Thames terminal there. The Kempsford, Cricklade and Lechlade houses survive today, as do the five round watchmen's houses that are a unique feature of this canal.

Several wharf houses survive on the Herefordshire & Gloucestershire Canal. The canal company owned twelve wharves in all on its 34-mile (54km) line; at one of these, Withington Marsh, the wharfinger's name, William Bird, is still just discernible painted on the end wall of the house, and others can be found at Ledbury, Canon Frome, Staplow, Kymin and Crews Pitch. These houses were more likely to survive in remote areas; in more populated districts they were often demolished in favour of modern development. Nor is it

(Above) The wharf at Bradford-on-Avon, Kennet & Avon Canal

(Below) On the Herefordshire & Gloucestershire Canal towards the end of the nineteenth century

surprising that the small hand-operated cranes, part of the equipment of very many wharves up and down the country, have not survived at all apart from a few that have been specially preserved, for example at the head of the Driffield Canal in Yorkshire.

Many wharves were served by horse-drawn tramroads that conveyed principally coal but also other bulk items such as iron, lime and sand from mines or quarries within a few miles of the canal. The Brecknock & Abergavenny Canal was supplied by something like fourteen tramroads, including the Hay Railway which extended 24 miles (38km) to Hay-on-Wye with a branch to Eardisley. There were tramroad connections to most of the South Wales canals; and on the 9¼ mile (15km) Penydarren tramroad connecting with the Glamorganshire Canal at Abercynon, Richard Trevethick's locomotive made its first journey in 1804. The Montgomeryshire Canal in mid-Wales carried lime, brought to it by tramroad at Pant and at Llanymynych where self-acting inclined planes were used to work the wagons down the steep gradient. And the garden of the Wharf House at Mamble on the Leominster Canal still yields fragments of edge rail, relics from the tramroad that brought coal from mines in the hills above. Finally the Caldon and Peak Forest canals were among many others served by tramroad connections.

Although they developed in the seventeenth century, the great age for tramroads was the first half of the nineteenth. Many of them were absorbed into railways but a few carried on operating until the century's closing years. In hilly country where railways did not take over their line some of the routes can still be traced, above the Brecknock & Abergavenny Canal near Llanfoist, for example, or in South Devon where you may find stone rails of the Haytor Granite Tramway that connected with the Stover Canal. They were essential components of what developed as an integrated local transport system, and some of the canal/tramroad junction wharves became centres of prosperity for the surrounding area. Here, warehouses and cranes would have been found, and a general air of bustling activity even in the depths of the countryside.

Nevertheless, compared to the great wharves and basins of the cities, the country canal wharves were like wayside railway stations or remote and isolated halts. While the number of wharves on a canal was a good indication as to its prosperity, it was its city wharves that really mattered as far as the shareholders were concerned. Of the Kennet & Avon's nineteen wharves, six were in Bath; without these the trade would never have been enough to have kept the canal going for more than a few months. In the cities, many wharves were owned or leased by large undertakings and factories and ran themselves, the canal company receiving income without being involved in any expenditure. Returns from most country wharves were miniscule by comparison.

Without documentary evidence it is not easy to reconstruct the daily life of the country wharf. There are, of course, records of tonnages and tolls, and the names of traders and the goods they handled, but nothing about the daily routine and the various happenings that would have enlivened it. With large quantities of goods held in store, security was a constant problem and a conscientious wharfinger must have spent many uneasy nights when his warehouses were full of items in local demand. Coal was always wanted, not only by the boatmen who traditionally never bought any, but by everyone. There are stories too of smuggling; brandy brought in from the Continent, transferred to canal boat at Brentford dock and carried up the Grand Junction to the Midlands for sale and distribution – Fradley Junction was a favoured place. With the fluctuating revenues of many country canals, wharfingers and toll-collectors often found it difficult to raise enough cash for disbursements, and frequently failed to collect the rents and charges due to them. When private owners who had leased wharves fell on hard times the canal company suffered as well, and private wharves often lay idle for long periods while new lessees were sought.

As well as wharves leased by private owners from the canal company there were also wharves in private hands by right. An eighteenth-century Canal Act

empowered the Lord of any Manor and the owner of any lands through which the canal should be made to erect and use any wharf, quay, landing-places, or warehouses in or upon their lands, adjoining or near to the canal, and to land any goods or other things upon such wharf, or upon the banks lying between the same and the canal. . .

and a subsequent judgement held that 'an owner of land adjoining the towing-path had a right to erect a wharf on his own soil, and to land goods on the towing-path, and convey them across it to his own wharf.'

Private ownership of wharves largely contributed to the demise of the Buckingham arm of the Grand Junction Canal (now part of the

Sleaford Navigation.

An ACCOUNT or SCHEDULE of the feveral GOODS, WARES, and MERCHANDIZES, which are to be taken and confidered as a TON, and to pay TOLL accordingly.

QUALITY.	QUANTITY.	TONS.	QUALITY.	QUANTITY.	TONS.
Coals, - -	1 Chaldron, -	1	1 hatch Reed,	5 Hundred,	1
Oats, - -	10 Quarter, -	1	Grocery, —	2 Hogfheads	1
Barley, -	6 Quarter, -	1	Latten Reed,	250 Bunches,	1
Malt, - -	10 Quarter, -	1	Soap, —	2 Hogfheads,	1
Wheat, -	5 Quarter, -	1	Woad, —	1 Hogfhead,	3 qrs.
Beans, -	5 Quarter, -	1	Spetches, -	8 Packs, —	1
Peas, - -	5 Quarter, -	1	Squares at 9 Inches	250 -	1
Rape, - -	5 Quarter, , -	1	Sheep, —	Twenty,	1
Bark, -	10 Quarter, -	1 .	Porter, -	6 Barrels,	1
Whole Lime, -	1 Chald. & half	1	Flour, —	8 Sacks, —	1
Sleck'd Lime, -	2 Chaldron,	1	Seed, - —	5 Quarter,	1
Potatoes, - -	130 Pecks, —	1	Hay, —	20 Hundred, —	1
Lime Stone, -	4 Hogfheads,	1	Glafs, - —	7 Whole Crates	1
Timber, (Oak, Afh, & Elm)	40 Feet, —	1	Hemp Seed, Pofts 4 and half	40 Strike, — 120,	1 1
Fir Timber, -	50 Feet, —	1	Coak, —	100 Strike, —	1
Bricks, -	5 Hundred, —	1	Pavements,	3 Hundred,	1
Flat Tile, -	1 Thoufand,	1	Stone, -	16 Feet Cubic,	1
Pan Tile, -	5 Hundred, —	1	Paving Stone,	10 Superf. Yards.	1
Oil Cakes, about 6lb. and half a Pair. Larger Cakes in Proportion,	1 Thoufand,	1 & half			
Wine, - -	2 Pipes, —	1			
Felloes, - -	120 -	1			
Seven Feet Pofts,	Sixty —	1			
Six Feet, Five Feet & half	Eighty, Ninety,	1			
Single Deals, -	Half Hundred	1			
Double Deals,	QuarterHund.	1			
Battens, -	1 Hundred	1			

N. B. ALL other Articles not mentioned in the foregoing Lift to be fubject to 2s. per Ton of 2240 Pounds, to be afcertained either by Weighing or Draught of Water.

B. CHEALES,

Clerk to the Company.

Sleaford Navigation toll charges

Grand Union Canal). Private owners acquired a monopoly of the trade, and the canal company, as well as other would-be traders, found themselves shut out of Buckingham trade altogether. Consequently the canal company's motivation for keeping the branch in good condition diminished and it became more and more difficult for boats to navigate the full length. In 1910 the top section to Buckingham was abandoned, although a few miles of the branch lingered on until 1938. It may still be possible to find the sites of wharves at Deanshanger and Leckhampstead, and Wharf Lane at Old Stratford leads to the site of Old Stratford Wharf where large brick-built vaults were built underground for storage. The boat-building firm of Hayes & Co, famous for its elegant launches for overseas buyers, also used this site during Victorian and Edwardian times.

Opened in 1801, the Buckingham branch, 10½ miles (17km) long with two locks, left the Grand Junction at Cosgrove and ran through gently rolling countryside close to the upper reaches of the Great Ouse. Much of its course may still be traced and the explorer

Re-cutting a section of the Buckingham arm of the Grand Junction Canal, Deanshanger, 1902. The benching method used would have scarcely changed since the eighteenth century (Waterways Museum)

can reflect on what has sadly been lost. One of the Buckinghamshire locks has been immortalised in an attractive painting by Frederick William Watts.

The perils of trading on the Buckinghamshire branch in its run-down condition in the early twentieth century is exemplified in the story of the lost cargo of granite. In July 1906 the surveyor to Buckinghamshire Rural District Council, Mr Varney, ordered 25 tons of granite from Mountsorrel on the Leicester Navigation to be delivered to the small country wharf at Thornton. At the end of the month he complained that the invoice had been received but not the granite. Enquiries to the carriers, Faulkners of Leighton Buzzard, proved fruitless. The granite company then instituted a search of all the wharves on the branch and eventually discovered that the cargo had been delivered to Deanshanger Wharf by mistake. The boat captain, it transpired, had failed to ascertain the whereabouts of the wharves on the branch, and another captain had told them to carry on until he saw a previous load of granite on a wharf and to deliver it there.

'He came to Deanshanger and saw some granite on the side and said, "This is my place", and shot it off. Then the Deanshanger wharfinger signed his back note without looking to see where the cargo was consigned, and away he went, and the cargo was lost, and nobody knew who or where anybody was.'

When the cargo was eventually discovered the owner of Thornton Wharf refused to have it delivered there as it would do too much damage, and the next wharf, Maids Moreton, was unreachable as the canal was 'foul'. Thus ended the Mountsorrel granite trade to Buckingham.

Trade elsewhere was also affected by problems at wharves. When a cargo of stone was offloaded, it helped the carter who had to remove it from the wharf to heap it up as high as possible, because unless the wharf was properly paved it was difficult to use a shovel on its surface – the convention was that one ton of stone should cover no more than one square yard. For the boatman, however, it was often advisable to offload from bow, midships and stern in alternation to prevent the boat grounding where the canal was shallow; though the wharfinger was not expected to sign for a delivery until he was satisfied that it was in order. Nevertheless, in an argument about a bad delivery of stone at Hickling Wharf, the wharfinger said that 'he did not care whether he got the cargo delivered there or not, and he would not risk having his head broken in an argument with a

Lock, lock cottage and pub: Mountsorrel on the Leicester Navigation. An oil painting by W. E. Cooke, c1880

bargee.' The 'bargee' in question was Captain Mellor, a time expired Highlander who had fought at Magersfontein – so the wharfinger may have had a point!

Stone delivered to country wharves for road building was a considerable trade in the East Midlands but their business was handicapped by the bureaucracy involved. A consignment from Mountsorrel to Hose on the Grantham Canal needed nine consignment notes, the boats were gauged four times and four accounts had to be settled. Canal carriers would not accept small consignments, either, and traders often found it difficult to ascertain in advance the charges for haulage, tolls and wharfage. There were other difficulties too, especially with the 'bargees'; the manager of the Mountsorrel Company put it like this:

> The owner of a couple of narrow boats and his crew are at a great disadvantage when pitted against such competitors as, shall we say, Mr Evans, the goods manager of the Midland Railway. Although I believe they are improved as a class, many of them are rough diamonds, at times they are light-fingered, they are not invariably sober, their education leaves much to

be desired, and it is impossible to allow them the run of the works after closing hours. They are not always on the best of terms with the surveyors and carters with whom they have to deal, and threats of actual violence do not tend to the increase of business.

The strongest complaint against the canals was of bad management. While the railways had advanced with the times the canals had not, even though many of them were railway-owned by the early twentieth century. Although, for example, one railway-owned canal, the Trent & Mersey, had the reputation of being the best managed in England, the poor condition of the connecting waterways hindered its trade. Lack of investment and the fact that over the years the canal companies had failed to get together and rationalise their affairs made them slow competitors with the more progressive railway companies. Unpaved country wharves without cranes where heavy material had to be shovelled upwards, firstly by the boatman onto the wharf and then by the carrier onto his cart, could not compare with railway stations where in one simple operation the carrier shovelled the stone or coal downward onto his cart and was away.

Few people led such an isolated life as some of the lock-keepers on country canals. Nowadays you rarely meet a lock-keeper except at busy and complicated flights of locks such as Foxton or Bingley, and most of the lock cottages built by canal companies have been demolished. It was not the number of locks that dictated the number of lock-keepers but the distance they were apart. On the Somersetshire Coal Canal two lock-keepers were employed to work boats through the twenty-two locks of the Combe Hay flight, whereas on the Wey & Arun line, with twenty-three locks, fifteen lock-keepers were needed. In the years of its prosperity, the Kennet & Avon Canal required forty-two lock-keepers out of a total staff of 122, including masons, labourers, carpenters, puddlers, two blacksmiths and two ballasters. The Herefordshire & Gloucestershire Canal Company, a much smaller concern, had twenty-three men on its payroll in 1876, including nine lock-keepers, to look after its 34-mile (54km) line. And among the Thames & Severn Canal's employees were twelve watchmen who also acted as lock-keepers for the locks within their length.

All lock-keepers had other responsibilities and worked long hours, sometimes throughout the night. Each was responsible for the maintenance of sluices and weirs within his length, for water levels and the condition of embankments; he might also have to act as toll-keeper and to supervise the gauging of boats. A lock-keeper paid rent for his cottage – on the Herefordshire & Gloucestershire Canal this was 2s a week out of a pay packet of 14s. Cottages varied greatly, some reflecting the wealth and status of the canal company, others the idiosyncracies or taste of their designers. There are hand some Telford-designed houses on the main line of the Shropshire Union Canal, and strange barrel-roofed bungalows on the Southern Stratford. The Gloucester & Berkeley Canal provided its bridge-keepers with miniature Doric temples as homes; the round houses on the Thames & Severn were more distinctive than convenient to live in. The stark up-and-down cottages on the Herefordshire & Gloucestershire contrast sharply with the picturesque half-timbering in the villages nearby; there are solid grey stone dwellings by the locks on the Caledonian Canal, and severe red-brick houses on most of the Midland narrow canals.

Almost all these cottages had gardens which would have been assiduously cultivated by the lock-keeper and his wife, and many had smallholdings attached to them. They compared well with the industrial terraced housing of the period although many of them were damp and liable to flooding. However, it must have been hard on the nerves of a mother of young children to live with a 75ft (23m) long trough, 15ft (4m) deep and often full of water, just a few steps outside her front door. The cottage by Somerton Deep Lock, isolated in the fields by the Southern Oxford Canal, was an especially eerie and hazardous place to live.

Some lock cottages in their canalside setting were extremely picturesque. John Hollingshead was delighted with those he saw on the country stretches of the Grand Junction during his voyage from London to Birmingham in 1858:

Many of the lock houses are very pretty. All of them are neat and clean. In some of the more important lock houses, the keeper is seated in a little counting-house amongst his books and papers; in some of the smaller ones, rude accounts are kept in mysterious chalk signs upon the doorway or the walls. . . At all the lock houses, coy little gardens peep out, and many of them are profusely decorated with flowers both inside and outside. One cottage on the canal bank, connected with the canal traffic, is such a complete nosegay that the word

'Office' and the City arms painted over its doorway, are scarcely visible for roses.

In her autobiography *Lock Keeper's Daughter* (Shepperton Swan, 1986), Pat Warner gives a vivid picture of her early years in the 1920s and '30s when she lived in Reservoir Cottage on the Tardebigge flight of locks on the Worcester & Birmingham Canal. Her father looked after fifteen locks and the reservoir that supplied them, as well as the hedges and ditches and the grass alongside the towing-path. On Sundays in the fishing season he had the additional task of walking or cycling several miles to sell permits for the Fishery Board. For all this he earned rather less than £2 a week.

Pat recalls one regular occurrence to which the nearness of the canal added an extra dimension:

To get to our lavatory on a winter's evening was like preparing to climb Mount Everest. You were made to dress up in warm clothes and undress when you arrived. Inside the small brick building with its squeaky door were two wooden seats. A big one and a little one. Toilet paper was only for the very rich. The

A lock cottage on the Droitwich Canal

daily newspaper and Old Moore's Almanack would be tied on a string and hung on the back of the door. As you sat on the wooden 'throne' the wind would howl through the trees and the water would lap against the wall at the back of the house. Similarly, the waves in the canal would splash against the coping stones along the edge of the lock at the front. Now and again, the owls would hoot and a vixen would scream. All just a little bit frightening when you are not yet five years of age.

Until the present century, lock-keepers' wages ranged from 10s 6d (52½p) a week, as paid by the Kennet Navigation and adjudged insufficient to live on, to about 25s (£1.25). On Ireland's Grand Canal the fifty-six lock-keepers each received only 9s (45p) a week in the early days; nevertheless the post was much in demand as the free house and garden was a considerable temptation. Wages in Ireland remained very low; the lock-keepers on the Barrow Navigation were earning only 11s (55p) a week as recently as 1930. Moreover the locks had to remain open all night. Ruth Delany in *Ireland's Inland Waterways* tells of Thomas Murphy, the Lowtown lock-keeper – when asked by the company chairman how long he had worked for them, he replied 'A hundred years, fifty by day and fifty by night'. Even so, the job of lock-keeper was often handed down from father to son, in Ireland especially, because secure employment with house and garden was a greater attraction than the wages.

Not only were wages in Ireland low, but lock-keepers could forfeit much of them in fines if they fell foul of the regulations. According to the rules of the Royal Canal in 1813, a lock-keeper would be fined 2s 6d (12½p) if he were absent from his lock without leave or if he failed to have his lock full with the gates open when the packet boat for Dublin arrived. A 5s (25p) fine would be levied if he left his lock full for two or three hours or if he let the racks fall without winding them down. And if he allowed a boat through his lock without a regular pass he would be 'dismissed from his station and rendered incapable of ever again serving upon the Line as a lock-keeper.' Royal Canal lock-keepers received 10s (50p) a week at this time, but did also have a free house and garden.

Lock-keepers on Scotland's Caledonian Canal fared rather better financially but then their work was of a different order. They had to be carpenters as well as lock-keepers and had one or two assistants – more at flights of locks such as Banavie – who were usually also craftsmen of some kind. They were paid about £1 a week and most of them had free houses and gardens. Some were able to rent small farms and one, James Rhodes, also ran an inn at Banavie for a time until his hostile attitude towards those skippers who avoided his inn led to his dismissal. Until the locks were mechanised in the 1960s the work was heavy, with two men needed to turn the capstans on each side. Earlier this century there were twelve lock-keepers employed at Banavie and eight at Fort Augustus; now there are just two at those flights.

Many lock-keepers' jobs were taken by retired boatmen looking for security in their later years in a world which they knew. Others spent their whole working lives 'on the bank', while several locks and lock cottages were passed on from father to son. No-one had a more intimate knowledge of the canal than a lock-keeper whose working life was inextricably tied to the level and movement of the water and the timber and gear of the locks and sluices under his control. Unhappily there are few of them left today and you may spend a week or more cruising on the canals without ever meeting a lock-keeper. Many of the lock cottages which have escaped demolition have been sold into private hands and have lost their connection with the canal.

Lengthsmen were employed by canal companies to look after sections of canal, especially on longer pounds where no lock-keepers were available. They were provided with accommodation and patrolled a few miles of waterway in either direction, keeping the towing paths clear of obstructions, laying and maintaining hedges and generally tending the track and its banks and embankments, culverts and drains. Theirs must have been the loneliest life of all canal workers, with conversation restricted to the few seconds of exchange with the crews of passing boats. Their successors, few as they are, live on housing estates in the nearby towns or villages and travel to work by bicycle or car, with possibly ten times as many miles of canal within their care.

Variety came to the lengthsman's life in times of crisis: he might be called on to join a gang of workers from the company yard on urgent repairs or to help with the operation of the spoon dredger. In frozen conditions in winter he might have to help crew an ice boat along with fourteen or fifteen others drawn from the company's labour force. The ice-boat crew would stand facing each other on either side of a central bar, rocking the boat energetically from side to side, while a team of ten or twelve horses, mostly borrowed from boats iced up on the canal and each with its own line and driver, charged ahead with as

A Cambridgeshire mole-catcher with his dog (Cambridgeshire Collection)

much power as they could muster. The combined efforts of tugging horses and rocking men chipped away at the ice, but seldom did so much effort have such little effect. Carried away by enthusiasm it was possible to tip the boat right over, which was bad luck especially on those far from home.

In addition to the wharfingers, lock-keepers and lengthsmen, a canal company also employed carpenters, blacksmiths, masons, bricklayers and labourers who moved to different sections of canal as occasion required. Many companies also employed a vermin killer or mole-catcher as rats, rabbits and moles could do serious damage to canal banks, tunnelling through and causing leaks and sometimes the eventual collapse of whole sections of embankment. Nevertheless, valuable though he was, the mole-catcher was not among the better paid employees; for example when times on the Brecknock & Abergavenny Canal were hard in the early 1820s, the mole-catcher's pay was reduced to £16 a year, although presumably he could still benefit from the sale of moleskins. However, not even the most skilled vermin killer or mole-catcher could deal with the problem on the Wendover arm of the Grand Junction Canal which was resulting in serious leakage. This was caused by 'a peculiar little animal which bred in very large quantities, just at the outcrop of the Totternhoe Stone [through which this branch was cut], and these continually pierced the puddle and created leaks.' This little animal was identified by the famous geologist Frank Buckland as a variety of small crab, and apart from reconstructing the canal with vertical walls and bitumen sheeting in the bottom, which was beyond their financial resources, there was nothing the company could do which would prevent its depredations.

Waterways in the past seem to have bred a special breed of men; or perhaps, like railways in pre-nationalised times, they attracted those for whom the occupation was not just a job but a whole way of life. 'I found life – real life – on the canals,' said one retired canal worker who in his time had been lengthsman, bricklayer, member of a dredging gang and lock-keeper. Many families were loyal to the waterways through successive generations; typical of these was the Siddons family – grandfather John, son George and grandson John served the River Nene Commissioners for a total of one hundred years, often taking reductions in wages when the Navigation fell on hard times, as happened frequently. Three generations of the Smale family kept the Werrington inclined plane on the Bude Canal. More recently, however, as commercial trading on the waterways has collapsed, the old family loyalties have come to an end and only reminiscences are left – and fewer and fewer of these as the years pass by.

2 Canalscape

The impact of a canal on the landscape in the late eighteenth/early nineteenth century can be compared with the impact of a motorway today. True, whereas the motorway sweeps confidently across the contours, the canal, especially the early one, dodged round them. But motorways, canals and railways as well, affect the landscape considerably. Initially almost everyone reacts in a similar way: it will be a monstrosity, a permanent eyesore, destroying the views, disrupting nature and bringing murder and mayhem to the settlements along the route. Yet as time passes and the raw wounds heal the monstrosity becomes familiar, and after a few years it is impossible to imagine the landscape without it. Motorways, railways and canals become refuges for wildlife, long cross-country corridors where flowers, birds and insects flourish that elsewhere may be killed by pesticides

The Kennet & Avon Canal, near Crofton

or predators. This is especially true of the country canal – whatever its promoters and engineers intended, it always enhances the scene and enriches the countryside through which it flows.

The engineers and builders of the early canals seemed to have an instinctive sympathy with the landscape. Yet their first objective was to get the work finished as quickly and cheaply as possible. To this end, stone from the locality was used in bridges and locks, or bricks might be made close to the site if there were no brickworks nearby. There is an unpretentious honesty about much of this early work; no-one, it seems, was out to make a name for himself by leaving a monument to his own skill or ability upon the landscape. The Staffordshire & Worcestershire and the Trent & Mersey canals are good examples of this unpretentiousness where the various structures – the locks, cottages and bridges – have an essential rightness about them; it seems inconceivable that they could have been designed or built in any other way.

Another noted example from rather later in time is the Brecknock & Abergavenny Canal, surveyed by Thomas Dadford and opened in 1812. This canal made an end-on connection with the earlier Monmouthshire Canal, and its prime purpose was to convey coal and iron from mines and ironworks in the Welsh hills, brought to the canalside by tramroad, on to the Monmouthshire Canal and thence to Newport. In its conception it was wholly industrial. The two major structures on the canal are the massive masonry Brynich Aqueduct over the Usk and the great embankment over the Clydach valley, 300ft long and 75ft high (91 × 23m), with the River Clydach carried through in a long tunnel. Both aqueduct and embankment now have a timeless quality about them as if they had always been there, and the canal itself seems as much a part of the natural scene as the River Usk below.

The profession of civil engineer grew to maturity during the canal age, and in the early part of the nineteenth century some of the leading figures were able to impose their personalities on the works that they designed. Such an opportunity came to John Rennie, engineer of the Kennet & Avon Canal, the 'stupendous national concern' providing 'a direct water communication between Bristol, Bath and the Metropolis'. Rennie was not impressed by the quality of the local stone, especially as it was being supplied before it was properly seasoned, and suggested using bricks instead; however, he was overruled by the canal committee and was obliged to use stone in the construction of the two major aqueducts, Dundas

Cherry Eye Bridge on the Caldon Canal in winter

Dundas Aqueduct on the Kennet & Avon Canal

and Avoncliffe. Dundas especially is an architectural masterpiece, reflecting the confidence of its designer and of the canal company in their undertaking. Classical in style, with a 4ft (1.2m) projecting cornice and elegant balustrades, for westward-bound traffic it is a fitting introduction to the splendours of Bath. London-bound travellers, however, see the reason for Rennie's misgivings; in sad contrast, that side of the aqueduct, exposed to the cold north winds, is patched and pitted. The same can be seen at Avoncliffe, a few miles to the east.

A few years earlier Rennie had worked on the Lancaster Canal, a section of which was opened in 1797. This included a splendid five-arch aqueduct over the River Lune, 60ft high and 600ft long (18 × 182m). Rennie wanted to use brick here as well, but again he was overruled. Although more utilitarian in design, the Lune Aqueduct also has a projecting cornice and lengths of balustrades, and is equally appropriate to Lancaster. Many of the other aqueducts and bridges on this canal have an architectural formality characteristic of John Rennie at this time.

Although perhaps no other canal engineer expressed himself as forcefully as Rennie, a few have left their own characteristic mark. Benjamin Outram's great aqueduct at Marple took the Peak Forest Canal across the River Goyt, and was described by George Borrow as 'the grand work of England', filling his mind with wonder when a boy. It is a handsome masonry structure with massive pillars of red sandstone set in the river beneath. The roundels between the arches, however, were not made for decorative purposes but to reduce the weight, and although the aqueduct was a tourist attraction for a time it was never intended as such.

The canal engineer who altered the landscape more than any other was Thomas Telford. While Brindley and his immediate successors built narrow canals that ambled quietly around the contours, with major engineering works kept to a minimum, Telford's canals strode through the countryside taking the shortest and most direct route. The new main line of the Birmingham Canal and the great cuttings and embankments on the main line of the Shropshire Union are notable examples.

It is in Scotland, however, that Telford's achievements are most striking. In the Lowlands there are the three major aqueducts on the Edinburgh & Glasgow Union Canal, at Slateford over the Water of Leith, across the Almond and, mightiest of all, the 810ft (246m) twelve-arch crossing of the River Avon. All three are very similar

Chirk Aqueduct. From an early engraving by T. Barber after a drawing by
H. Gastineau

Laggan cutting on the Caledonian Canal

Pontcysyllte Aqueduct. An early interpretation from a drawing by G. Pickering

in construction and design to Telford's aqueduct at Chirk on the Llangollen Canal, all of masonry with similar iron troughs and iron railings. From the Avon aqueduct in 1823 'the woody glens, the rugged heights, and the beautiful Alpine scenery around, must raise sensations of pleasure in every feeling heart'.

However, Telford's principal contribution to the landscape is in the Highlands. In 1803 he was appointed engineer to the proposed Caledonian Canal, with William Jessop, with whom he had often worked, as consultant – though Jessop died eight years before the canal was completed. The Caledonian, 60 miles (96km) long with locks able to accommodate vessels of 160ft by 36ft (48 × 10m), was built at a cost of a little less than one million pounds, and was the greatest canal undertaking in Britain and probably the largest civil engineering enterprise up to that time. The Caledonian took nineteen years to construct, and at the busiest periods almost a thousand men were employed at any one time. The large-scale works, such as the flights of locks and the Laggan cutting, altered the landscape and might have scarred it for ever, had it not been for Telford's insistence that some half-million trees and shrubs be planted alongside, both to stabilise the banks and to improve the visual impact.

Telford was surveyor or engineer for various other waterways, including the new Birmingham main line. He made the survey for

A typical fenland waterway

Stretham Old Engine on the Old West River in the Fens

the Macclesfield Canal, although the engineering of this splendid route was the work of William Crossley. But he is best known for the tall and elegant aqueduct at Pontcysyllte on what was originally the Ellesmere Canal, now known as the Llangollen. The original proposal for this 'wonderful Edifice', as the canal committee chairman described it, was William Jessop's, principal engineer to the canal company; however, the main credit for the design and execution seems to belong to Telford.

The canal engineer whose work showed the closest sympathy with the landscape – melding with it rather than imposing upon it – was the first of them all, James Brindley. He surveyed the routes of thirteen canals and acted as engineer for eight of them, although in many instances he left the completion of the task to others. Usually Brindley's canals fitted the contours of the land, though when difficult problems arose he could always find a solution – sometimes this was novel, even radical, such as the Barton Aqueduct on the Bridgwater Canal, 'a river hung in the air', 'the greatest artificial curiosity in the world'; or the Harecastle Tunnel, at first derided as impossible but later described as 'our eighth wonder of the world'.

Brindley's approach to canal building has been neatly expressed by Samuel Smiles. 'He would rather go round an obstacle in the shape of an elevated range of country than go through it, especially if in going round and avoiding expense he could accommodate a number of towns and villages.' His canals therefore did much to help the development of the smaller settlements on their routes – for example the Staffordshire & Worcestershire, the Grand Trunk (later to become the Trent & Mersey) and the Chesterfield Canal. As far as possible Brindley avoided major engineering works and used local materials in the construction of bridges and locks, and so his canals are an integral part of their environment.

By contrast, there is one area of Britain where past waterway construction, instead of integrating with or enhancing the landscape, radically altered it so that its whole character was changed. In the fenlands of Cambridgeshire and South Lincolnshire waterway construction took place many years before the age of canals had dawned: the landscape is a direct consequence of the vast fen drainage scheme carried out under the direction of Cornelius Vermuyden in the mid-seventeenth century.

The medieval fenland was a vast area of marsh and quagmire, with islands rising out of it here and there on which religious settlements were founded. Closest to the Wash was a band of silt fen; elsewhere the main constituent was peat. Through this boggy land, haunt of wildfowl, eels and frogs, rivers meandered seawards, sluggish in flow and their estuaries often silting up, their channels indeterminate and frequently shifting. These rivers were the Nene, Welland and the Great Ouse with its tributaries Cam, Lark, Little Ouse, Wissey and Nar; together with waterways now extinct, notably the Wellstream, they were the main, and in many districts the only, means of transport. Building stone from the quarries at Barnack, a few miles west of Peterborough, was carried by boat to the sites of the cathedrals, abbeys and many of the churches in Lincolnshire and Cambridgeshire, and to Cambridge for the colleges of the developing university. Items for daily use which could not be obtained or manufactured on the spot were brought in from the larger towns on the periphery of the fenland.

These waterways supplied the great fairs at Sturbridge, near Cambridge, St Ives and Reach with both merchandise and merchants from France, Italy, Spain and the Baltic countries. Boats were also used for passenger transport within the Fens and for carrying the region's own produce, especially reeds, rushes, peat and fish. So, even though these fenlands were subject to frequent flooding both by the sea and by upland water, there was much local opposition to proposals to drain them for fear that navigation might be impeded and the wealth of the area then pass into the hands of outsiders.

From time to time over the centuries there had been efforts to drain areas of the Fens, usually by individual landowners, but what generally happened was that after some years the effort and cost of maintaining the drainage proved too much and matters reverted to what they had been. With the dissolution of the monasteries under Henry VIII things became worse since, dispossessed, the wealthy religious establishments no longer improved the condition of their holdings. In 1588 a team of surveyors advocated large-scale drainage by carrying off the water in straight canals to the sea, and predicted that a great increase in profit would accrue to the Crown. However, nothing was to happen for many years.

Further enquiries and proposals for drainage took place in the early seventeenth century, but it was not until 1630 that a group of landowners led by Francis, Earl of Bedford, agreed to undertake to drain the southern fenland in return for grants of tracts of the drained land. The Dutchman Cornelius Vermuyden was appointed engineer in charge and work soon began. Between 1630 and 1637 six straight drainage channels were constructed, including the 21-mile (33km)

Bedford River, sluices were inserted and old channels improved. This, however, proved inadequate, and further operations were halted because of the Civil War. After the conflict Vermuyden was reappointed to complete the work; several new cuts were made, one of them the New Bedford or Hundred Foot River to supplement what became the Old Bedford and create a reservoir for floodwater between them. In March 1653 commissioners inspected the works and adjudged the Fens fully drained – an optimistic judgement, as it turned out.

Many alterations and improvements have been made to the fen drainage system since 1653, including elements originally proposed by Vermuyden but not undertaken then because of insufficient funds. Nevertheless the basic pattern of the system is his, and it

The fenland waterways before Vermuyden

is his engineering that is primarily responsible for today's fenland landscape. Vermuyden's concern was with drainage, but most of his larger channels were – and still are – fully navigable. Until the 1930s such channels as the New Bedford, the lower reaches of the Old Bedford, the Forty Foot, the Sixteen Foot and Bevill's Leam were used by gangs of lighters to supply the fen farms and villages and take their agricultural produce, as well as to carry coal and later oil to the many pumping stations. Today the occasional pleasure cruiser may be sighted on these waterways but most of the time they are deserted except for flocks of fishermen at weekends.

Vermuyden's theory was a simple one: to carry off the surplus water from the rivers and streams by a series of straight cuts, some leading into larger cuts such as the Bedford rivers, and thence by the shortest possible route to the sea. An unnatural pattern was thus imposed on the countryside, as can be seen from a glance at the map. There was also a major and unforeseen consequence that affected the landscape in another significant way: as the land was drained, the peat dried out and began to shrink and waste away; so the surface level sank and this affected the drainage channels. Professor H. C. Darby describes it thus:

> Within a comparatively short time from the completion of the reclamation, many of the main channels of the Bedford Level flowed at a higher level than their subsidiary feeders. These in turn were higher than the minor drains. And these minor drains, in their turn, became increasingly shallower and ineffective. This extraordinary paradox can be seen along the fen rivers of today. Along the New Bedford River, for example, the height of the barrier bank is greater from the outside than from the bed of the river.

Vermuyden had planned that drainage would be by gravity but this became impossible. The solution was to pump the water from lower to higher level drains; first of all this was by wind pumps, of which at one time there were several hundred, then by coal-fired pumping engines, then diesel-powered engines – now it is by electrically operated pumping stations. The steeply raised banks of the drains added another feature to the landscape.

So as you cruise along these straight canals today, what do you see? On some routes nothing much, as the high banks effectively cut off the view and there is nothing except the water before and behind and the sky above. If you moor and climb to the top of the bank the flat fields stretch out before you, the horizon broken only by a few trees here and there, perhaps a church spire, perhaps if it is very clear the west tower and octagon of Ely Cathedral. You may become aware of the islands rising out of the fen: the Isle of Ely, and the smaller islands where the settlements of Chatteris and Littleport have developed. But to obtain some idea of the changes wrought upon the landscape by Vermuyden and his successors you have to go to Wicken Fen.

Now owned by the National Trust, Wicken Fen has survived as an undrained area of fenland and is now several feet higher than the drained lands surrounding it. Wicken is a sedge fen and is not typical of the old fenland; also it has been managed for centuries, as indeed it still is. But here, and looking across the adjacent Adventurers' Fen, drained in the past but now dedicated to the interests of wildlife, you may be able to recapture something of the feeling of things as they were before the drainers got to work; and before the arable farmers took over, too, grubbing up the hedgerows and sowing their crops, creating vast fields which stretch out to an artificial horizon formed by the nearest raised canal banks – the New Bedford, or the canalised Ten Mile River – all the product of man's ambitious engineering.

The drainers, then, remodelled the landscape and the canal builders enriched it. In places the canals have blended into the countryside so closely that it is easy to mistake them for natural rivers; the upper reaches of the lockless Ashby Canal, for example, or the many miles of the long summit level of the Leicester arm of the Grand Union. Professor W. G. Hoskins paid eloquent tribute to the overall contribution of the canal builders:

> The Canal Mania, as it was called, produced some beautiful landscapes, chiefly in the Midlands and the North where the new large-scale industry needed cheap transport; they were important visually too because in the Midlands above all large sheets of natural water were uncommon; no coasts, no lakes, only ponds and streams. And the canals brought their own special kind of buildings: wharves, corn-mills as at Shardlow, warehouses, quays for unloading coal in remote

(Opposite) A lock on the Brecknock & Abergavenny Canal (British Waterways Board)

country districts, lovely Sheraton-bowed redbrick bridges of hand-made brick, soft and shining in the sun, towpaths, inns, and lock-keepers' cottages. It was a whole new world, which jaded town-dwellers in overcrowded England are beginning to discover for themselves.

All this places a responsibility on those who must repair and maintain the canal system today, or who earn their living from supplying the needs of canal users. However, visual impact cannot usually be taken into account, nor can materials such as Professor Hoskins' 'hand-made brick, soft and shining in the sun' be used when repairing or rebuilding bridges that have to carry today's heavy motor traffic. Too often the solution adopted has had to be the cheapest and easiest; a patching-up job with inappropriate modern materials or a pre-stressed concrete construction thrown across for road-users. There *are* instances where exceptional care has been taken – the new bridge by the Coach and Horses near Llangynidr bottom lock on the Brecknock & Abergavenny Canal, for example, is in modern idiom with a rectangular bridge-hole but is both functional and decorative; but until quite recently the twentieth century has seen little to be proud of in canal architecture. We seem to have lost the almost instinctive sensitivity that all past builders seemed to possess, men as diverse as James Brindley, William Jessop, John Rennie, Thomas Telford, the Dadfords father and son, Robert Whitworth and James Green.

Even late in the period of canal building this sensitivity had not disappeared. The Macclesfield Canal was opened in 1831, more than seventy years after the opening of the St Helen's and the Bridgewater canals, at a time when railways were beginning to grip the minds of would-be promoters and the general public. Telford surveyed the route and laid out a line as direct as possible between the Trent & Mersey Canal at Kidsgrove and the Peak Forest Canal at Marple. High embankments and deep cuttings were particular features and all twelve locks, which raised the canal 118 feet (36m) to a height of more than 500 feet (152m) above sea level, were grouped together at Bosley. The detailed work on the canal was the responsibility of William Crosley, the engineer, and the many fine stone bridges were built to his design. The Bosley flight of locks is one of the most beautifully sited in England, and the canal as a whole, throughout the 28 miles (45km) of its length, blends perfectly with the countryside and the towns through which it runs.

Lockside dwellings on the Grand Union Canal

Only recently, in the last thirty years or so, have people in general become aware of the beauty of the country canalscape. In 1944 L. T. C. Rolt began his account of canal travels, *Narrow Boat*, with the words:

> Most people know no more of the canals than they do of the old green roads which the pack-horse trains once travelled. . . Knowledge of them is confined to the narrow hump-backed bridges which trap the incautious motorist, or to an occasional glimpse from the train of a ribbon of still water winding through the meadows to some unknown destination.

More than any other book, Rolt's *Narrow Boat* was responsible for redirecting attention to the attractions of canals; but was it in fact redirection? Over thirty years before Rolt, E. Temple Thurston, intent upon discovery and wondering where to go, came to a decision:

> 'I'll go,' said I one day, 'where are no guides and scarce a map is printed. Who knows his way about the canals of England?'
> 'They begin at Regent's Park,' said a man.

'And then?' I asked him.

'There's one passes near Slough on the Great Western. I've seen it from the train.'

'If that's all that's known about them,' said I, 'I'll get a barge myself and go on till I stop.'

Fifty-three years earlier in *Household Words* Charles Dickens published John Hollingshead's account of a voyage on the Grand Junction Canal from London to Birmingham. 'I was an extraordinary tourist and my point of starting was a Basin,' he said. His cabman had never before been ordered to a barge-wharf by the side of a basin, and everyone he met on the voyage found it hard to understand what he was doing and why he was interested. His own 'long cherished notions of the dry, utilitarian character of canals' soon disappeared 'to give place to a feeling of admiration for the picturesque beauty of the country, and the artificial river, lying and running unheeded so near the metropolis.' 'Unheeded' is the key word.

Back in 1819, however, thirty-nine years earlier still, the Grand Junction Canal was a tourist attraction and the subject of the colour-plate book *A Tour of the Grand Junction Canal* by John Hassell, dedicated to its proprietors. The canal is described as:

> an almost perpetual succession of variegated beauty, shaping its devious course through some of the richest valleys of Middlesex, Hertfordshire, Buckinghamshire and Northamp-tonshire accompanied by a redundance of the most luxuriant scenery, and lined on its sides with a succession of rising eminences.

Hassell undertook his tour on horseback, visiting various stately homes on the way, but he considered the canal to be of at least equal importance for his readers. By this time the Grand Junction, apart from the long tunnel at Blisworth, had been open along the main line for nineteen years. The rawness had long worn off and the waterway with its locks and bridges had become an integral part of the rural scene, as can be seen from Hassell's aquatints.

The scenic attractions of another canal had been extolled in the dignified columns of *The Times* ten years before Hassell's *Tour*. The maiden voyage on this canal took place in October 1809 when the proprietors' vessel led a procession of loaded barges belonging to local traders, all of whom 'hoisted flags or streamers and whatever should testify their joy that all their speculations of a profitable traffic were now realised.' As the fleet passed through a forest, so the report continued:

> The proprietors found their calculations of profit irresistibly interrupted by the rich prospects breaking upon them from time to time by openings among the trees, and as they passed along they were deprived of this grand scenery only by another and no less gratification, that of finding themselves gliding through the deepest recesses of the forest, where nothing met the eye but the elegant windings of the clear and still canal, its borders adorned by a profusion of trees of which the beauty was heightened by the tints of autumn.

The canal which boasted this sylvan setting and which also provided wide views over the fields and hills of Kent and Surrey was the Croydon Canal, a 9¼ mile (15km) water link between New Cross and Croydon. Unfortunately, despite the hopes of its proprietors and its popularity with sightseers, pleasure boaters, anglers and bathers, the canal was a commercial failure. In 1846 it was sold to the London & Croydon Railway Company, which closed the canal and built its line over much of the route. Tracing the canal today, through Brockley, Honor Oak, Forest Hill, Sydenham, Penge, Anerley and South Norwood to West Croydon station, is a challenge to the imagination. Nevertheless, in its day the Croydon Canal opened up the countryside to dwellers in both South London and Croydon – for them, it was genuinely a country canal.

The year 1846 was important in canal history, for no fewer than seventeen navigations came under railway control. When John Hassell undertook his tour of the Grand Junction in 1819, the only railways that existed – horse-drawn or powered – acted as supply lines to canals; there were only a few isolated instances of competition, such as that between the Croydon Canal itself and the Surrey Iron Railway which took goods to Croydon. The early railways were all local concerns and it was not until the 1840s that the full possibilities of this new mode of transport were realised involving routes on a national scale. This precipitated the season of railway mania, with 179 Railway Acts being passed in three years and about one-fifth of inland navigations passing into railway control.

For the canals it was the beginning of the end. They were out-of-date, slow, unexciting and inefficient. They still performed a useful function, and were still indispensable in regions where railways had not yet penetrated, but now they lacked both novelty value and glamour – both of which the railways possessed. Their glories were of the past. 'The work of Titans rather than the production of our Pigmy race of beings' Josiah Wedgwood had written of the Runcorn locks on the Bridgewater Canal; and when in July 1788 King George III and his Queen had visited Sapperton tunnel, then under construction on the Thames & Severn Canal, they had expressed 'the most decided astonishment and commendation' at the work. But by the middle of the nineteenth century those days were over and the canals in general were unheeded, except by the traders who still used them to transport their goods, or the boatmen and their families who earned their living by them. Harold Schofield, reporting on his waterway journey from Manchester to London in 1869, described the Thames & Severn Canal as 'one of the most beautiful we ever were on, and the water is the purest I have seen,' but added 'the canal boats were conspicuous only by their absence.'

One small group of people, however, now turned to the canals for inspiration. In the earlier years of the Canal Age, artists had shown little interest, apart from occasional bread-and-butter paintings and

The Kennet & Avon Canal, near Newbury. The sketch is by John Constable, June 1821

sketches of aqueducts for engraving. But as industries developed and smoke blackened the skies, painters began to look to the country canals as characteristic of a rural England under threat – a timeless countryside where water trickled through old lock gates, boatmen relaxed at the tiller while their horses plodded the overgrown towing-path, and children fished for minnows or dangled their fingers idly in the stream.

The greatest of these painters was John Constable, whose paintings of the Stour navigation show his feeling for boats and boat-building, watermills and locks – he wrote himself '. . . painting is another word for feeling'. In June 1821 Constable accompanied his friend, Archdeacon John Fisher, on a tour of Berkshire and Oxford-shire during which he made several sketches of the Kennet & Avon Canal near Newbury. However, it is the artists of lesser rank who have treated the subject in a more idealised or sentimentalised way, that have left us the 'prettiest pictures' of nineteenth-century canals.

One such was Frederick William Watts, himself much influenced by Constable; his oil paintings 'The Lock Gate' and 'An Old Lock in Buckinghamshire' both depict a peaceful canal scene, the boatmen, with all the time in the world, easing their boats into the lock while the horses quietly crop the grass and the summer day rolls by. Then there is Benjamin Leader, the son of Leader Williams who was designer and engineer of the Anderton Lift on the Trent & Mersey Canal; Benjamin was widely known for his canal and river scenes, many of which were engraved for the enjoyment of a larger public. Miles Birket Foster was also attracted to waterside subjects which he depicted in a gentle, elegiac style.

However, the continuing importance of the canal as a transport route was advertised with éclat when, on 2 October 1874, 5 tons of gunpowder being carried by fly-boat exploded beneath Macclesfield Bridge in Regent's Park. According to the leader writer of the *Illustrated London News*:

A destructive force equal to, if not exceeding, the most violent of tropical tornadoes was, by some cause not yet ascertained, suddenly set free to expend itself in a few brief seconds in distributing within a wide circle death, havoc and ruin. It was as swift, as devastating, as irresistible, as a bolt from heaven.

Three people were killed: the steerer of the *Tilbury* which had the gunpowder on board, a crew member and a boy. The zoo animals were shaken up, too – 'The elands and antelopes, the giraffes, the elephants, and a rhinoceros, showed great excitement.' The disaster prompted the magazine to tell its readers about canals in general, and to include several wood engravings showing scenes from normal canal life as well as representations of the calamity. The courses of both the Regent's Canal and the Grand Junction were described, an article on canal-boat life was published, and tea-time on a monkey boat was illustrated (see page 52).

Now that regular commercial traffic on the narrow canals has ended and the old boatmen and their wives are dead or retired the canal world is no longer closed or remote. For most people the canals are now holiday playgrounds, popular in the warmer months, cruise-ways for pleasure cruisers. Today's visitor wants a well-dredged channel, a firm towing-path, locks that fill and empty quickly, plenty of water points and sanitary stations. Large hiring boatyards with extensive car-parking space have grown up on the sites of many old wharves. The old way is enshrined in museums; the new way with its amenities, transistor radios, daily papers and prettily painted boats is captured on video camera to be enjoyed again on winter evenings. The picturesque element beloved by the Victorian painters is less in evidence; crumbling red-brick bridges are condemned by Highways Departments as unsafe for vehicular traffic and are replaced too often by harsh concrete structures, towing-paths are repaired with hard straight edges, and lock machinery is 'modernised'. Stretches may sometimes be found where time seems to have stood still, but there are fewer of them every year.

3 Life on Board

'This was a new mode of travelling and a delightful one it proved,' wrote Don Mañuel Alvarez Espriella in one of his *Letters from England* in 1807. Don Mañuel, in reality the poet and essayist Robert Southey, was enjoying a tour of the North West and had embarked in a 'stage boat' bound for Chester.

> The shape of the machine resembles the common representation of Noah's Ark, except that the roof is flatter, so made for the convenience of passengers. Within this floating house are two apartments, seats in which are hired at different prices, the parlour and the kitchen. Two horses, harnessed one before the other, tow it along at the rate of a league an hour; the very pace which it is pleasant to keep up with when walking on the bank. The canal is just wide enough for two boats to pass; sometimes we sprung ashore, sometimes stood or sat upon the roof – till to our surprise we were called down to dinner, and found that as good a meal had been prepared in the back part of the boat while we were going on, as would have been supplied at an inn. We joined in a wish that the same kind of travelling were extended everywhere; no time was lost; kitchen and cellar travelled with us; the motion was imperceptible, we could neither be overturned nor run away with; if we sunk there was not depth of water to drown us; we could read as conveniently as in a house, or sleep as quietly as in a bed.

This was canal travelling in style, the waterway equivalent in its time of the first-class Pullman car on the old 'Brighton Belle'. In the hard commercial world, however, things were very different, and became more so as the century progressed. In *Canal Adventures by Moonlight* – not a book to seek out if you are looking for a romantic thriller – George Smith, the reformer, describes a six-day voyage in a monkey boat from London to Leicester in September 1880. Equipped with carpet bags full of bull's eyes, tracts and copies of the *Christian Herald*, he boarded the *Ouse* at Paddington Stop. The crew consisted of Captain T——, his two sons and a featherless blackbird,

'On the Road' featured in *The Illustrated London News* October 1874

and the cargo of rice, leather, paraffin, hemp-seed and oil. Jerry, the motive power, had served over eleven years on the towing-path and had walked an estimated 80,000 miles (128,744km).

On the first day Smith recorded passing a boat loaded with 20 tons of rotten eggs, and also observed 'two big, strong, coarse, burly, half-nude either boat-women or tramps' who had been washing their clothes and themselves in the canal. 'This sort of disgraceful procedure was going on in broad daylight, within the sound of the church bells of the capital of Christendom, and the passing and re-passing of canal boats upon which were herds of men, women and children of both sexes and all ages.' Near Uxbridge the black sludge of the canal gave way to clear water with fish darting about and pebbles visible at the bottom; a group of 'coarse naked lads' were frisking about in the water and Smith threw bull's eyes for them to dive after. They stopped at Rickmansworth for the night; there was no room for Smith to sleep on board so he found lodgings at the canal side where he supped off 'rice pudden' and ginger beer, which gave him nightmares.

The second day began with a fight on the Birmingham boat

which was accompanying them – subsequently one of the crew was forced to seek refuge on *Ouse*, to Smith's further discomfort. But the passage through Lady Keppel's Park near Watford was very beautiful, with 'lovely green fields, rippling spring-water brooks, bounding deer, rich foliage of the hedgerows, stately trees, scents of wild flowers, charming scenery, picturesque views and delightful weather', spoiled only by passing boats on which there were 'lots of swearing women and children, as ignorant as Hottentots'. Smith's day was made by hearing one of the boat lads on the Birmingham boat singing a Sunday School Union hymn as he rode the horse along the towing-path.

After a night at Leighton Buzzard Smith settled down to observe the traffic on the canal; this included boats from the Oakthorpe, Moira and Warwickshire coalfields, gas-tar boats from Birmingham and boats laden with salt from Droitwich. Where there were children on board he threw bull's eyes to them. A boat from Staffordshire laden with general merchandise met with his approval; it was registered and painted, its crew clean and orderly and it passed by noiselessly. By contrast came a steam tug, worked by a pregnant woman 'who seemed to be captain, engineer, stoker, in fact everything'; this one confirmed Smith in his intention to have the Canal Boats Act brought into full force.

In the early evening *Ouse* arrived at Blisworth Tunnel where a steam tug soon emerged to draw them through, a noisy and uncomfortable process enlivened by the singing of songs and hymns by the woman on the Birmingham boat. After passing Heyford Ironworks they eventually tied up at Buckby Wharf at 1.30am, having covered some 40 miles (64km) that day.

The next day being Sunday *Ouse* remained at her moorings and Smith went to chapel in Long Buckby; in the afternoon he handed out sweets and tracts to the local boat children. At 4.30am they were off again, taking the Leicester arm, or the Grand Union ditch, as the captain called it. Passage along the narrow cut was impeded by weeds and bulrushes but the going improved on the far side of Kilworth Tunnel. Here, the canal

. . . seemed to be a union of all that was lovely and enchanting, and not, as in other places, a union of poverty, misery and wretchedness. On the sides there were overhanging pine trees, which were covered with masses of foliage of various hues and tints. The hedges were covered with honeysuckle and blackberries. . . The water was covered with a sheet of green moss, very much resembling a sheet of green ice, through which boats, water-rats, water-hens and fish had left their trail, presenting an appearance in the distance of one huge black and green carpet. The chirping and whistling of Charlie, the crack of Little Matt's whip, the tramping and snorting of Jerry, the conversation of Mr T, echoing and blending together in the distance over this watery grave, made it most delightful on this hot summer day.

That night Smith had to sleep among the bags of rice in the hold, which occasioned a series of dreams that took him thirty pages to describe – or rather gave him an opportunity to recount imaginary conversations and encounters to reinforce his campaign for full implementation of the Canal Boats Act. They moved off at five o'clock next morning on the final run to Leicester, descending twenty-four locks, many of them 'deep and tumble-down'. Word had gone ahead that Smith was on board a Midland Counties Carrying Company's boat, and several people came out to have a look at the well-known reformer whose efforts on behalf of the boat children,

'An Evening Halt' featured in *The Illustrated London News* October 1874

although not always appreciated, were widely known. He left the boat at the company's warehouse, distributed the remaining bull's eyes and tracts, and hurried off to prepare his next onslaught on the correspondence columns of the press based on the experiences of his voyage.

Smith's books and his letters to the newspapers serve as a corrective to the romanticised view of canal life seen in the paintings of, for example, Benjamin Leader, or in the pages of more modern writers who nostalgically recall some golden age. At times Smith obviously overstates his case, especially with regard to numbers: he records far more family boats than others on the canal system, which is firmly contradicted by evidence from the Census. Nevertheless his case was a strong one; many of the boats *were* no better than filthy floating slums, crewed by an illiterate couple with perhaps as many as eight young children, drawn by a worn-out nag and carrying a cargo of stinking tar-water or even rotten eggs. And many a miserable infant was drowned in the cold waters of the cut.

However, this was not the whole story, for at the time Smith made his journey, and throughout the whole period he was investigating the lives and conditions of the boat families, stressing their poverty, fecklessness and dirtiness, several potteries in the Church Gresley area were busy manufacturing the so-called 'bargee' wares, largely intended for purchase by the boat people themselves. These were of heavy pottery with a dark brown glaze and decorated with applied sprigs of flowers and birds, usually peacocks, and started to appear in the 1860s. Enormous wide-bellied teapots, often with a miniature teapot in place of a knob on top of the lid, were the most popular – many were sold at a shop in Measham on the Ashby Canal, hence the alternative name of 'Measham Ware'. A special characteristic was the motto, an applied label bearing a legend impressed with printers' letters, such as 'A present to a friend' or 'Welcome Home'. A personal motto could be ordered in advance and collected on the next visit to the shop; hence many of these wares carry the owner's name and address or are personalised gifts, such as 'From his Mother to Percy Henry Robinson'.

Although the large teapots were impractical for daily use in a crowded narrow-boat cabin, they were popular as wedding presents or to mark an anniversary. There was a touch of prestige in owning such a splendid article and the common descriptive term 'bargee ware' or 'barge teapots' testifies to the close association of this ware with canal people. This would indicate that not all boat families

'Tea-time on a Monkey Boat' featured in *The Illustrated London News*, October 1874

were destitute and that there were still plenty who took a pride in their homes on board and who were eager to ensure that they were properly equipped.

Another fashion which began in the 1880s was the large-scale importation from the Continent of cheap porcelain pictorial souvenirs, and many boat people soon became collectors. Plates with pierced edges carrying an illustration of a famous building or seaside town with 'A Present from. . . .' in Gothic script were very popular; others illustrated royal occasions such as jubilees, coronations and weddings, and yet more bore pictures of flowers, some hand-coloured or hand-finished with gilding around the borders. With ribbon threaded through the piercing these plates could be hung around the cabin walls providing cheap and bright decoration, the candle or lamplight reflecting off them cheerfully in the evenings. Some cabins might have had twenty or so on display, with as many more stored beneath the seats.

Yet an article in the *Leicester Daily Post* at about this time, inspired by George Smith's earlier attempt to enforce the Canal

Boats Act, said of the boat families: 'They lead a merely animal existence. Their chief amusement is to get drunk. . . they represent almost as low a type of life as we still have in England, and there are 100,000 of them.' Another newspaper, the *Derby Mercury*, asked:

> Do our readers know what sort of beings comprise our canal population? It may be taken for granted that they are human, because they can swear and get drunk. . . Swearing and drunkenness seem to be their two ruling passions. Public houses of the lowest kind are along the sides of the canals. There the men, women and children are to be seen at all hours of the day, and fighting, ruffianism, blackguardism of the worst kind are indulged in.

There is much more of this sort of thing which seems designed both to horrify and terrify the newspapers' readers.

This is in direct contrast to the boat people who were proudly equipping their cabins with Measham ware and hanging plates. Some of the women would also crochet lace draperies for the cabin walls, and even little hats for the horses to wear to keep the flies from their ears. Moreover the census returns for the 1870s and 80s indicate that there were no more than 40,000 people involved in navigating the inland waterways, with fewer than 9,000 making their homes permanently on boats. There is no denying that some boatmen drank, and at times drank too much, but in that respect they were little different from other members of the 'labouring classes'. According to the registrar general's report for 1890–2, the number of boatmen who died from alcoholism was the same as carmen and carriers, and considerably fewer than seamen and dock workers, although still rather higher than the national average.

From these contradictory accounts what does become clear is that conditions varied enormously, and that what was observed on one section of the canal system was often very different from what was seen elsewhere. George Smith saw overcrowded family boats carrying cargoes such as refuse or gas-water, poorly maintained and towed by broken-down nags, but then he attacked and condemned the way of life of virtually *all* users of the canals – although he did grudgingly admit that he had also found 'many kind-hearted, clean and respectable people'. And certainly not all boatmen could afford costly Measham teapots or days off to go to fairs or seaside resorts to buy fairings to decorate their cabins, nor would they even think of doing so.

There is, however, no doubt that the conditions of boat people did improve in the period between the passing of the Canal Boats Act in 1877 and the beginning of World War I. Furthermore, the amended Act of 1884, which made inspection of boats compulsory, had even more effect. George Smith deserves much of the credit for these Acts; they were very much the result of his lobbying and letter-writing, generally stirring the conscience of both the public and members of parliament. The Boatmen's Friend Society also helped considerably; it provided for the practical needs of boat people – for example it set up attractive rest rooms as an alternative to the public house – and because of their unsanctimonious, friendly approach its missionaries often became welcome guests aboard and were valued for their help. Moreover, throughout the nation there was a general improvement in standards of health – better surgical techniques and nursing therefore enhanced the prospects of recovery for boatmen who were notoriously accident-prone.

Was there ever a Golden Age of the canals? If there was, no-one at the time records having lived through it, but that would be true of all Golden Ages. The years following World War I must have been

'The Lock' featured in *The Illustrated London News* October 1874

one of the better periods, however; if only for a short time, the canals were enjoying a measure of prosperity while the railways struggled to make good the shortage of rolling stock and lack of investment in maintenance as a result of the war. Boatmen of the 1950s and '60s looked back on their younger days as the good times; times of hard work and tough competition, but good times nevertheless, when the whole family lived and worked together sharing a way of life that was handed down through two or three generations and had become a tradition. That tradition, however, was short-lived and past figures for the annual total tonnage carried on the canals tell the story:

1913	31½ million tons
1918	21½ million tons
1924	17 million tons
1938	13 million tons
1946	10 million tons

In the following years the decline continued. The canals were nationalised and in a crass attempt to impose an appearance of unity on the waterways, narrow boats painted in a new livery of blue and yellow appeared. For a time some trade survived: sugar to Bournville, lime juice on the southern Grand Union, imported grain to Wellingborough on the River Nene via the Grand Union and the Northampton arm, and various other isolated runs. Often Blue Line Canal Carriers worked seven days a week carrying coal to the jam factory at Southall, but that trade ended in 1970. In the following years a few individuals tried their hand at commercial trading, but regular long-term contracts were no more – on the narrow canals and the country canals generally trading was finished.

Reminiscences, however, abounded; during the 1970s Jack James, retired lock-keeper and boatman whose own collection of canal artefacts inspired the foundation of the Waterways Museum, used to hold court at the Boat at Stoke Bruerne, recounting tales of the golden years on the Grand Union and the Oxford canals – and who is to blame him if the gold glittered more brightly as the eager listeners clustered round and the pint glasses steadily accumulated? A wonderful BBC archive recording of 1969 preserves the voices and recollections of several of the old boatmen, lock-keepers and toll-collectors, including George Bate, Joe and Rose Skinner, Ernie Thomas, 'Chocolate' Charlie Atkins, Sam Lomas and Jack James himself. There are the stories of Sam Lomas's 'rheumatic' telephone, of Billy Biggins who changed clothes with a scarecrow who he thought was better dressed than he was, of the boatman who fought for every one of the twenty-four locks at Chester, and of the social evenings at the pub.

Within less than three decades a whole way of life has passed away. Nearly all the old boatmen are now dead, and the evidence of their past – their world – is preserved only in museums and old photographs. It would be foolish to romanticise or glamourise this world as their work was hard and their hours long; sometimes from 5am until 9 or 10pm at night, in all seasons and weathers, with none of the domestic comforts or the living and working space that we take for granted. Yet at the same time we must accept – and perhaps even envy – the claim of one old boatman, looking back on his life:

> Dark nights come, and that's the dreariest part of the season and the dreariest part of the canal is the winter season. The water looks real murky. But we get over it. We look forward to the spring again, and the birds singing, and the cuckoo back with us. On the canal, we live with Nature, and Nature lives with us. And we love it.

On the Hereford & Gloucester Canal; after a painting by Philip Ballard (see page 147)

4 The Transformation of a Canal Village

It was a fine spring day in the year 1768 and the stranger had stayed at the inn overnight. He had retired to bed early and soon after dawn had called for his mare to be saddled up. Several people noticed him as he rode through the village in the early morning, along the High Street, down Mill Lane, then turning to head east through the fields close to the river. He had seemed deep in concentration.

That evening the stranger's visit was commented on by the handful of villagers gathered in the Red Lion. Those who had heard him speak tried to place his origin from his accent; most felt that he was from somewhere up north. His clothes were workmanlike, his mare in good condition but old. However, in no way did he resemble other strangers that visited the village from time to time – he had shown no interest in buying or selling anything; he had asked for no individual either by name or trade and had left behind him neither samples nor notices. Someone suggested that he might be a spy from France but the impression of solid Englishness that he had created made this seem ridiculous.

Molly, however, when she came in to clear away the mugs and tankards had something to offer. If the talk was about him who had stayed in number one last night then he'd been drawing on the floor by the bed. She couldn't make out what it was so she'd left it for someone else to come and see.

At first the drawings, in white chalk, made no sense at all. They were just lines, with small squares and triangles here and there, and arrows scattered along them. Not until Tom sat on the bed and looked down on them did things begin to fall into place. It seemed to be a plan of part of the village. That could be the church – there was the High Street and the Red Lion. Mill Lane was there, and the mill, and the wiggly line must be the river. Perhaps the spy theory wasn't so silly after all. But there was one line that didn't fit in with any local features, which ran between the village and the river; then just past the mill it stopped. At the point nearest to the mill two short lines crossed it and there was a small smudgy rectangle on the village side.

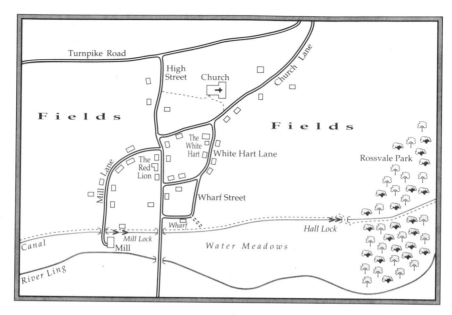

The village shortly after the opening of the canal

Downstairs the general opinion was that the drawing represented the line of a new road, possibly a turnpike, though there seemed no reason for choosing that particular route except that it kept to flat ground. Only the miller, John Blake, seemed likely to be directly affected by it as the line ran along the borders of his property, and in places right over it. So Molly was sent to ask him to come and have a look; then at least he would have been forewarned.

It did not take Blake long to assess the chalk drawing, though at first he asked several questions about the stranger, especially about his bearing, the way he spoke and about his horse. At last he seemed satisfied, and told them it was no new road that was coming to the village. No: it was something modern, something useful that might provide some well-paid jobs and could bring the price of coal down with a bump. He reckoned the stranger was James Brindley, a millwright from maybe Staffordshire or Derbyshire, well known as an expert at his job. This Brindley had been doing wonders in Manchester and other places, he'd heard, but not any more with mills. He would bring a new invention to the village, something very different. A canal.

By next midday the news was all round the village and folk were peering westward or down towards the river as if expecting the canal suddenly to materialise. But months passed, and nothing happened.

Perhaps John Blake was wrong, or perhaps the canal would go else-where. Expectations subsided for almost a year. Then suddenly they revived; one of the local farmers returned from a distant market with news that strange men had been seen pacing the fields and that an office had been set up in the White Hart on the market square with a notice in the window about a canal. There had been something in the local gazette about an Act of Parliament. There had also been talk in the market about raising money, and he'd been asked if he would like to buy shares in a canal company.

Again the village buzzed with conjecture; hopes were raised of well-paid jobs and cheaper coal. A major cleaning-up operation began at the Red Lion; beds were repaired, floors were scrubbed and the stable roof was mended. That Sunday the rector spoke of Mammon and temptation in his sermon, and on Monday he sent a letter to his bankers in the nearby town. Then a few weeks later news came from the squire's coachman. This man had taken Sir Richard to a ceremony on the outskirts of a town some 20 miles (32km) away where the first sod of the new canal had been cut by the county's lord lieutenant. There had been a banquet and the firing of several cannon, and hogsheads of ale had been distributed. It transpired that Sir Richard was to be on the canal company's committee, and it would not be long before the village would see some action.

By now people had taken time to think and not everyone was in favour of the new adventure. If the canal was a success then the packhorse trade would suffer and the carters would need to look elsewhere to keep themselves in business. And how would the canal operate? If the boats were horse-drawn it could benefit the blacksmith and the farmers who could provide fodder, but if they were to be sailed no-one would profit. Worst of all would be hauling by teams of men, the method apparently used at times on the rivers Severn and Thames – these bow-hauliers were said to be devastators and ravagers of the vilest kind.

However, many considered that John Blake would advise. He was known as a keen man of business and a master of his trade; further-more, he had travelled widely in the Midlands and the North and his views usually commanded respect. Some years ago he had led local opposition to a plan for making the river navigable, on the grounds that this would interfere with the supply of water to the mill and would therefore adversely affect both himself and the village. But the canal was different, and there was every chance that it would provide cheap transport for his flour to villages and towns on its route.

By the time an agent of the canal company came to stay at the Red Lion later in the week, most villagers had made up their minds about the forthcoming canal. A notice appeared in the inn window offering jobs to two masons, a smith and a carpenter, and an unspecified number of labourers. And other jobs would be on offer when the canal was completed if persons of the right quality presented themselves. Such promises of employment in an area where few lived at anything more than subsistence level were encouraging, and when the agent, Mr Morgan, appeared, several men were already waiting to see him.

Then a message arrived which caused considerable concern. The assistant engineer for this section of canal had fallen and broken his leg, and unless a replacement could be found the works would be delayed by several weeks. Morgan advised the men in front of him to seek other employment for the time being until the situation was resolved, when he would return. However, much was his surprise when a grey-haired but muscular workman called William Huckle asked him what an assistant engineer had to do. Morgan talked about lines, gradients and levels, water control, weirs and locks, and Huckle listened closely, seeming to understand. And when Morgan asked what he knew about all this, Huckle pointed towards the window:

'Those meadows down near the river,' he said; 'I did the watering of them some years ago.' This awoke Morgan's interest and the two men agreed to inspect the meadows as soon as it was light next day.

What Morgan saw was a group of four meadows, two on either side of a tributary stream. Weirs had been built in the stream at various points and above them were cut at right angles two straight drains, one on each side, with sluices at their head. Other drains led off these, also at approximately right angles, so that each meadow was intersected by a series of narrow parallel trenches. At the far end of each meadow drains connected with the stream so that surplus water could be returned.

Having shown Morgan the operation of the weirs and sluices, Huckle then took him to another meadow which was watered by a main taken off the river itself some distance away. The river was too wide for a weir to be erected across it so the main was supplied by the natural flow of the current. A system of weirs and sluices led the water from large trenches into smaller ones, with surplus water being led off into waste ground at the far end. Huckle explained that he had seen this method of watering meadows while on a visit to his

sister in South Cerney, Gloucestershire; he had modified it to water
his own meadows and had then been employed by Sir Richard who
owned the large meadow watered by the river.

'There's two of us here who knows about water,' he said to
Morgan. 'One's John Blake but he's always busy with the mill, and
the other's me. I'd come to see if you'd any use for me when I heard
about your engineer. I've not worked on canals, but then no-one has
round here. So maybe I'm your man.'

Two days later Huckle was asked to report to the canal company's offices as soon as he could, to meet the resident engineer
and accompany him along that part of the cut already completed.
Morgan reopened his office in the village and continued to recruit
his labourers, and Huckle's appointment as assistant engineer at a
wage of 25s (£1 25p) a week was soon officially confirmed.

Two months later four men arrived with measuring rods, chains,
stakes and a waywiser, and walked through the fields staking out
the line of the canal. Farmers were assured that where the canal
passed through their land, bridges would be built so that they and
their cattle could cross without difficulty. At the point on the canal
line nearest to the village the men marked out a large rectangular
area close to the track leading to a bridge over the river. They also
paused for some time a hundred yards further east where there was
a slight declivity, and again marked out a sizeable area. A quarter
of a mile further on, where the boundary wall of Lord Rossvale's
estate came down to the river, they stopped and then returned,
checking on the placing of their markers. Later that day people
trickled down to inspect the proposed line and to estimate if and
how it might affect them.

In the following week the village experienced the first benefits of
the new canal. Cutting had reached a point 8 miles (13km) to the
west where a lock was under construction. Water was let in as far as
a temporary dam at the head of the proposed lock and an area beside
the canal was levelled off to create a wharf. Here the first boatloads
of coal were unloaded and Sam Bridson, the carter, brought two
waggonloads to the village where it sold for 6d (2½p) a hundredweight, 2d (1p) less than it usually cost. Bridson also negotiated with
the canal company to put himself and his waggons at their disposal,
mainly for fetching and carrying construction materials; this would
compensate in some measure for the carting business he would lose
when the canal was completed.

The small group of canal labourers would leave their cottages an hour before dawn to walk to the workings where they would collect shovels to dig out the line below the half-completed lock. Now William Huckle could return home every evening, to sleep in his own bed instead of in lodgings. He had left his farm in the charge of his wife and son so that he could give his full attention to work on the canal, work which he was finding surprisingly satisfying – new problems arose daily and new solutions had to be found. In particular there were difficulties with the supply of water as the canal crept further and further from the newly made reservoirs in the hills. Water had to be taken off the river and here Huckle, with his experience of watering meadows, was quick to see the possibilities and to judge where channels should be cut and weirs and sluices installed.

Labourers from the villages along the line were by no means the only men digging out the cut – in the Red Lion stories were told about the wild Irishmen who worked like Trojans, ate like horses and, like King David of old, scattered their seed around the land. Some of those Irishmen would soon be working within a mile or two of the village. They would not be seeking lodgings as they slept in encampments under canvas, but all the same. . . New bolts were fitted to cottage doors, padlocks were fitted to henhouses, dogs were bought or borrowed from nearby farmers and the few marriageable daughters were sent off to stay with relatives in distant parts. The little brewery behind the Red Lion worked overtime and the baker ordered extra supplies of flour from John Blake.

However, there was a feeling of anti-climax when the navigators, as they were known, did eventually arrive; they were camped nearly 3 miles (5km) away. Each morning one of Sam Bridson's carters took bread, meat, ale, eggs and milk out to the camp, though some of the Irishmen did plod up to the village in the evenings and sat in the Red Lion, often grumbling about the foreman and the engineer who had decided on such a remote site for their camp.

The next few weeks were full of activity. As one group of navvies excavated the channel to a depth of 5ft (1.5m), others stacked and consolidated the spoil and began to set out the line of the towing path. Cartloads of clay for puddling arrived and were deposited at regular intervals along the line. The canal company had established its own brickworks some 10 miles (16km) to the west; boats conveyed the bricks along the watered length of canal and waggons brought them the final distance to the sites of bridges and locks. Masons were shaping the coping stones for the locks and carpenters were fashioning the weir doors and lock gates. A master bricklayer from

Staffordshire lodged in the Red Lion and rode down to the works early each day. On Fridays, Mr Morgan appeared to check on the amount of work done and to pay the craftsmen and labourers. Sometimes there was not enough money to pay them in full, which led to shouts, arguments and threats and occasionally to a workman storming out altogether.

Most active of all was William Huckle; he walked tens of miles each day, up and down the line of the canal, directing, improvising, sometimes seizing a shovel or saw to show exactly how it should be done. Often in the evenings he would walk to the mill to consult John Blake over a question of water levels or the exact placing of a sluice. Once a week the chief engineer appeared; he was a distant relation of James Brindley who had moved on to survey proposed canals in Birmingham and the Coventry area, and who was now very sick, suffering from diabetes and with little time to live. The chief was content to follow Brindley's line as far as possible and to leave as much as he could to his assistant.

By midsummer the lock to the west of the village was complete and water was admitted. Four bridges had been built for the farmers, and a substantial and handsome bridge carried the main carriage route to the south across the canal close to the mill. Work was in progress on the nearby Mill Lock, and a low brick wall was being erected along the boundary of the wharf. Below Mill Lock the channel was dug to a depth of 3ft (1m) as far as the boundary of Lord Rossvale's estate. The weather was fine and everything was on schedule.

Then one Monday morning when the men turned up for work they were received by Mr Morgan with unwelcome news; work was to stop immediately and all the local workers were to return home. The company had failed to come to an agreement with Lord Rossvale for the canal to pass through his land, and to make a diversion at this stage would mean two crossings of the river and an extra 2 miles (3km) of cut which they could not afford. The Irishmen would be moved to the far side of Rossvale's estate to begin digging there, while the company tried again to persuade his lordship to reduce his demand for £4,000 in compensation for land needed for the canal. Later, Morgan told Huckle that the Duke of Bridgewater had been forced to buy the whole Hulme Hall estate for £9,000 in order to complete his canal to Manchester, and Rossvale's land was worth four times that much – but anyway, he would never sell.

So activity ceased for almost a year; of the villagers, only William Huckle remained working on the canal. The Irish navvies dug another mile of cut and were then moved with Huckle to work on digging out wharves and basins at the proposed terminus some 30 miles (48km) to the east. Attempts continued to persuade Rossvale to accept the canal. It was not that he opposed canals on principle; in fact he already owned shares in a venture in Worcestershire. But he was inordinately proud of the landscaping of his park, improved by the great Capability Brown a few years before, and he did not want it bisected by an ugly ditch – nor did he want his game poached by rough and ignorant boatmen. Moreover the canal would cross his north carriage drive which would mean a crude red-brick bridge within sight of the Hall. Finally Sir Richard Pendlebury was a major shareholder in the company and Sir Richard was an old parliamentary opponent of his.

It was Sir Richard, concerned about his dividends, who proposed that the company call in Robert Whitworth as consultant engineer. Whitworth had been trained by Brindley and was acclaimed as a problem-solver in canal projects in the North. He studied the plans and obtained permission to walk through Rossvale's estate with Huckle, and together they worked out a proposal which he submitted to his lordship in person.

Lord Rossvale's stately bridge, repaired and restored in 1893

An early nineteenth century oil painting believed to be of Hall Lock

The proposal involved repositioning a lock, which would now be placed by the western boundary of the estate so the canal could enter Rossvale's grounds at a lower level, concealed from view in a cutting. A cross-over bridge by the lock would move the towing path to the far side of the canal from the Hall thus diminishing the possibility of poaching. The north carriage drive would be carried over the canal by a fine masonry bridge which would embellish the scene rather than detract from it. And it might be possible, if his lordship desired it, for a stretch of canal to be widened to form an ornamental lake visible from the bridge, where fountains and swans might add enchantment to the view. All this of course would involve the company in a great deal more expense but in the end it would be to everyone's benefit. Finally Sir Richard let it be known quietly that if the proposal was accepted he might consider withdrawing his support for a certain candidate in the forthcoming election, ensuring that Rossvale's nominee would have a clear run.

It was a tempting offer and Rossvale accepted it providing that the company paid £1,500 for the land, to which they reluctantly agreed. Whitworth drew up the plans and returned to Yorkshire; Morgan placed a notice in the *Gazette* and reopened his office in the Red Lion; Huckle moved back into his own home; two dozen of the Irishmen were recalled and in late July work recommenced.

By the end of the year the repositioned Hall Lock was completed and the navvies were at work in Rossvale's park. Now water could be admitted, and for a time the village became the terminus of the western section of the canal. A small brick warehouse and a cottage were erected on the wharf and another cottage on spare ground beside Mill Lock. Two permanent appointments were now on offer: a wharfinger, to be responsible for all goods passing through the wharf and able to keep records and accounts, and a lock-keeper, to look after Mill and Hall locks and 3 miles (5km) of canal with the various weirs and sluices. Thomas Rodd, once the parish clerk, became wharfinger and moved into the cottage with his wife and young son; and as lock-keeper Huckle proposed Joseph Tranter, who had assisted him with his meadow-watering undertaking. Tranter was sent to the western end of the canal to study lock operation and maintenance for a few days, and then the three traders already operating on the canal were told that their boats could come through.

There was little ceremony when the first boat arrived at the wharf, as no-one expected it. Rodd woke up late, looked out of the window and there it was, a long, narrow open vessel with a short, stubby mast. A man and a youth were standing by it and a horse was cropping the grass by the towing-path. Rodd pulled on his clothes, filled a jug with ale from a barrel in the kitchen, and ran down to greet them. Solemnly the three of them drank a toast to 'the success of the Navigation'. Mrs Rodd ran to the village with news of the boat's arrival and a small crowd soon gathered at the wharf eager to see it and discover the nature of its cargo. Mr Morgan and Sam Bridson arrived in one of Bridson's carts with an 18-gallon barrel of ale and some cans from the Red Lion. Morgan stood in the cart and delivered what he described as a few remarks but which Welsh fervour extended to half an hour. Eventually everyone drank to the success of the Navigation, and then half-a-dozen volunteers helped to unload the 5 tons of coal comprising the first cargo. Morgan showed Rodd how to deal with the paperwork and by the end of the day all the coal had been sold at 6d (2½p) a hundredweight, 2 tons going to Sir Richard's agent, 1 ton to John Blake and the rest in single or half hundredweights around the village. Two hundredweights were reserved for Tranter, who was entitled to free coal as a perquisite of his job.

Next day Morgan left, his task in the village completed, and a representative of the canal carriers arrived at the wharf to make arrangements with interested parties about a regular trade. There was a general demand for coal, and several farmers wanted a supply of lime for their fields. Bricks and timber for building were also needed and Sir Richard's agent, and the wealthier farmers' wives, asked for groceries, tea and spirits. As return carriage there would be produce for market and flour from Blake's mill as well as hay and straw in season. Until the canal was completed, the representative said, trade would be relatively small, but within a year or so the village would be a prosperous trading centre rivalling the market towns and there would be plenty of jobs and money for everyone.

In fact it was nearly three years before the canal was opened throughout. The cutting through Rossvale's grounds proved more difficult than anticipated; twice the sides collapsed and enormous quantities of clay were required to prevent leakage. The fine masonry bridge was not fine enough for his lordship and a master mason had to be brought from Bath, together with a supply of Bath stone, before he was satisfied. Elsewhere an aqueduct over the river had to be rebuilt and the workmanship on several of the locks proved faulty and had to be repaired. Then the canal company ran out of funds and was forced to obtain another Act of Parliament to raise more money; luckily other canal companies were doing well so there was no shortage of subscribers. They therefore felt justified in holding a Grand Celebration Banquet in the Guildhall of the county town; Lord Rossvale himself accepted an invitation and took the opportunity to order a new dinner service from the leading manufacturer of the day, Josiah Wedgwood, a guest of honour and himself a great supporter of canals. All joined in the chorus of a specially composed ballad:

Sing success to our Great Navigation,
The Pride of our Race and our Nation!
'Twill bring Wealth to all,
Both in Cottage and Hall,
The century's Greatest Creation!

and so on for several more verses. Hopes were high; prosperity seemed just around the corner.

5 The Canal Village
Through the Years

Several decades have passed. In the country generally there are signs of increased prosperity. Small towns, especially in the North and Midlands, have grown into large manufacturing centres. Enclosures have made agriculture more efficient although the condition of the peasantry has not much improved – many have moved to the towns in search of regular, if ill-paid, employment in the new works and factories.

Much of this prosperity is the result of the fully developed inland waterway transport system which now extends to a total of nearly 4,000 miles (6,437km) in Britain. Every large town has its canal connection. Especially busy are the canals around Birmingham, the routes across the Pennines, the canals serving the Potteries and the main line between London, the Midlands and the North. In South Wales canals have turned Cardiff, Newport and Swansea into busy ports exporting vast tonnages of coal and iron. In Scotland the Forth and the Clyde are linked by canal, and there is a sea-to-sea route through the Highlands along the Great Glen.

Nevertheless there are problems and difficulties. The narrow locks of the Midland canals restrict passage to boats of no more than 7ft (2m) beam, whereas other canals – notably the Grand Junction – can accommodate craft twice that width. Each canal company has its own regulations and table of tolls so long-haul traffic is sometimes delayed, and transhipment of cargoes necessary. Many canal companies have put dividends above investment, and neglect maintenance until an aqueduct or lock collapses and something *has* to be done. And there is much talk of railways; several miles are already in operation in the North and the risk of competition is obvious. Some canal proprietors choose to ignore this threat believing that they are powerful enough to block any proposal for a railway; others simply soldier on, intending if the railway approaches their territory to sell out to it for the best possible price when the time comes. Most canal companies, however, are content to carry on as they are while the threat is still at a distance, and while comfortable dividends are rolling in.

The canal has brought some changes to the village, but not the prosperity to the county that was promised. The terminal towns have expanded but not remarkably so, and the eastern terminus has always been a dead-end as the canal connection that was scheduled never materialised. Trade has therefore been mainly one way, from west to east, with little back carriage apart from seasonal agricultural produce. Coal, lime and building materials have remained the principal cargoes heading east, but the supply of coal has been diminishing lately owing to the greatly increased demand from the growing industries of the Midlands. The company is solvent, although Lord Rossvale's demands during construction meant a heavy burden of loans which took nearly twenty years to pay off. Most years a small dividend is paid but shares are not in any demand.

Developments subsequent to the canal include the wharf which has been paved and extended, two warehouses, a blacksmith's shop, a hand-operated crane and a small office adjacent to the wharfinger's house. Thomas Rodd and his wife have been dead now for many years, but their son Jason is wharfinger and he has an assistant, Joe Bridson, a grandson of the old carter. Bridson & Sons is a well-established business with several waggons and two boats on the canal, and has its own private wharf a hundred yards above Mill Lock with Joe's uncle Jason in charge. There is still a Tranter at the lock, too – Benjamin, Joseph's eldest son. Two houses have been built below the lock on the towing-path side; one belongs to the canal company and is rented to Billy Ricks, assistant lock-keeper and canal carpenter responsible for repairs to lock gates and sluices within 12 miles (19km). The other is a beer house, the Ship & Anchor.

The road up to the village has been somewhat improved, but the village seems to have changed little. However, the old thatched roofs of some cottages have been replaced by slate, and bricks of different colour and texture have been used in a few more recent buildings and in the repair of walls and outbuildings. Some of the boat captains now buy their clothes and boots in the village where the prices are lower than in the market town – so outside the bootmaker's are two pairs of massive hobnailed boots, and in the window of the village tailor's cottage is a fine red plush waistcoat.

Both the Red Lion and the church look much the same. The hopes of landlord and rector have not been realised; the Red Lion has remained an ordinary village public with two bedrooms for overnight guests, and St Peter's is still fifteenth century and earlier, as no new benefactor enriched by canal-borne prosperity has appeared to extend the chancel or rebuild the tower. Among the more recent graves in the churchyard, however, are those of John Blake and William Huckle. On the latter's headstone is a rustic inscription:

> Reader pause, and shed a tear,
> Here lies a worthy engineer.
> Water he tamed for good of all,
> Dear Lord receive his blessed soul.

After Huckle's death the whole line was placed under the surveillance of a superintendent based at the western terminus. Reports from the lock-keepers and toll-collectors now come to his office, and instructions are returned as to what should be done. Once a year he takes the directors of the company on a tour of the canal in the committee boat; this is usually the only occasion on which he is seen by the company's forty employees.

At this point there seems no reason for things to change. The canal has survived some difficult years, and unlike many of the waterways in southern and south-western England it has remained solvent and held on to its trade. The works are mostly in good condition and the staff in general are loyal and reliable. The towns and villages along the route depend on it for coal, and the weekly market boats have made life much easier for both buyers and sellers. The regular passenger service had to be withdrawn as it did not pay, but one of Bridson's boats occasionally takes passengers and there is talk of a through fly-boat service beginning next year. Altogether the canal is an essential part of life and it is impossible to imagine the village or indeed the county without it.

We move onward again, into the twentieth century. At first the canal scene seems much the same. A horse-drawn boat is locking through Mill Lock, its captain chatting with the lock-keeper. But the village wharf is in a sad state, one warehouse has partly collapsed and much of the wharf area is overgrown; the wharfinger's house is empty, too, with two of its windows broken. Only in the blacksmith's shop is there any activity: Rob Whiffle, the smith, is repairing the

(Opposite) The village seems to have changed little . . .

wheel of a bicycle. Blake's mill has disappeared, and in its place has arisen a large, gaunt, red-brick building with a tall chimney. A board proclaims it as Smerdon & Company's Flour Mill, and a small wharf has been made on the towing-path side close by – two boats are moored to it.

The main road is macadamised and there is a cast-iron, diamond-shaped notice by the bridge stating the weight limitations for locomotives and other vehicular traffic. Some of the old cottages in the village have vanished, replaced by grim-looking pairs of uniform red-brick dwellings. Half-a-dozen or so large villas have been built on the eastern side of the village; the largest has steep gables, a tiny turret and a monkey puzzle tree in the front garden, and is the home of George Bridson whose name is now emblazoned over a new workshop in the High Street with a petrol pump outside it. Here in the adjoining yard are several waggons, a van and an open lorry; Bridson was one of the first in the county to see the potential of the internal combustion engine, and he sold the firm's canal boats to help pay for the new vehicles.

St Peter's Church is not what it was. The old tower has been demolished and a tall steeple erected on its base; the chancel has been extended, and its windows filled with some disastrous stained glass from London's East End. Much of this was paid for by the squire, Sir Richard Pendlebury, grandson of the Sir Richard who had initially supported the canal. Astutely advised, he had sold the inherited family shareholding at the end of the year in which the canal company had shown its largest profit, this being mostly due to the transport of railway construction materials. When the improvements to St Peter's were completed, Sir Richard sold the manor to the governors of a newly founded public school, and departed to spend the rest of his life in warmer climes.

Until the sale of the family shares, there had always been a Pendlebury on the canal company's board of directors, a powerful voice in the local interest. After the sale and Sir Richard's departure, however, the company succumbed to the general uneasiness about the prospects for the future concerning canal undertakings generally. Negotiations were started with the West Midshires Railway Company which was contemplating extending its network to take in the canal's terminal towns. Purchase of the canal company would suit the railway excellently; much of the line of the canal would be suitable for the railway route and stations could be built on several of the wharf sites without having to buy yet more land. Matters were proceeding smoothly when there was an unexpected intervention. It came, ironically, from the present Lord Rossvale, whose great-uncle had caused the canal company so much extra expense over a century before. On no account would he allow the dirty, stinking railway to come anywhere near his property, and that included the many acres of his tenanted farms in the vicinity. Even bribery did not succeed, and so the railway company broke off negotiations – a few years later when the line was built, most of it lay about 2 miles (3km) north of the canal. However, the railway did buy most of the wharf and warehouse area of the eastern terminus where they built a fine station in the Gothic style. At this point in time the village is served by a halt at the end of a remote lane, but this is a long walk for housewives returning from market heavily laden. So the market boat still operates once a week, with an occasional extra excursion at weekends when it is sometimes used for Sunday School outings.

The coal traffic has moved to the railway as it proved cheaper and more convenient for loading, and Bridson has opened a coalyard by the halt. Other heavy traffic has followed, especially building materials and materials for road-making. Smerdons the millers, however, remain faithful to the canal which serves both the village mill and the two other mills they now own along its line. This is no sentimental attachment to water transport; it takes about a week for the imported grain to reach the mills from the port, and during that time Smerdons benefit from free storage on the narrow boats. With their own wharves there is no extra transport or handling charge and so, for the time being at least, the trade continues. Smerdons has a contract with the last surviving carrying company, and this has a continuing campaign against the canal company for more frequent dredging and improved maintenance of the locks. But apart from Smerdons' trade and the market boat, the carrying company only ever uses the canal for an occasional through-run of non-perishables for which there is no hurry in delivery. Some trade is carried on by a handful of family boats, owned and worked by the boatman and his wife and their older children; they will carry anything from fertiliser to household furniture and their brightly painted boats and gleaming brasses are attractive features of the canal scene.

These boat crews patronise the Ship & Anchor, as do many of the village labourers since the Red Lion smartened itself up to cater for the Pendle School parents. With its wooden settles, scrubbed tables, sawdusted floors and thin beer the Ship & Anchor will never attract the carriage trade but it suits the canal people. The lock-keeper –

another Tranter, Benjamin's son Simeon – spends much of his time here; the few boatmen that pass can work the locks themselves and visits from his superiors in the company are rare and always preceded by a message. Smerdons' workers come down most evenings and the Ship & Anchor is a lively place on pay-day.

The chief engineer of the canal company has recently been giving evidence before the Royal Commission on the Canals and Inland Navigations of the United Kingdom. Asked how he saw the future of canals, he agreed with other witnesses that the best prospects lay in companies amalgamating, or at least working very closely together, so that a uniform system of tolls and rates could be applied. A uniform gauge for all canals was also desirable to enable boats of, say, 9ft 6in (2.8m) beam to pass through every lock in the system – many boats were barred by the 7ft (2m) width of the narrower canal locks and bridgeholes. Where the money for this would come from he could not say, unless it was provided by the government.

When he returned the company began talks with others in the region, and preliminary plans for a union have now been drawn up; the Midshire Union Canal would have a total length of 142 miles (227km) and incorporate four companies. The feeling is that if the amalgamation comes about the future of canals in the region will be secure, despite the railways and the possible threat from the roads which more and more lorries are using every month.

The Commission, however, was not very impressed by the chief engineer. His answers to questions about the operation of his own canal were vague, and he seemed to have little experience of handling the day-to-day problems. Lord Shuttleworth, chairman, remarked to Philip Snowden MP, member, that he hoped this chief engineer was not typical, and that his subordinates knew rather more about the running of the canal than he did. In this he was correct: Thomas Dodwell was assistant engineer at £250 a year, and had learned his trade on the heavily locked Worcester & Birmingham Canal; it was Dodwell who travelled the line regularly and attended to all the problems and difficulties that arose.

The following decades see a continuing slow decline in the fortunes of the canal. The report of the Royal Commission recommended the widening and major improvement of many waterway routes and the nationalisation of some of them – but it is ignored, despite the estimated savings to canal users that would accrue if this happened. On the outbreak of war those canals in railway ownership are put under State control, but the independent waterways are left to themselves. Many of the younger canal workers leave to join the army, but somehow our particular undertaking survives; it takes some irregular traffic in agricultural produce and supplies and occasional loads of building materials for the new munitions' factories in the eastern terminus.

A few years after the war, with trade continuing to diminish and losing money every year, the canal company seeks to obtain a parliamentary Act of Abandonment which would lead to the closure of the canal and the sale of its assets. However, it meets with strong opposition – from the carrying company, from Smerdons (now a national milling concern and a contributor to majority party funds), and from the present Lord Rossvale who enjoys the picturesque element the canal contributes to his estate. So the directors agree to soldier on until an Abandonment Act can be obtained. In two or three years, the chairman thinks, Smerdons will yield to the attractions of road transport and their opposition will be withdrawn.

But Smerdons are still using the canal at the outbreak of World War II, and there is government pressure for more traffic to use the waterways. The carrying company has now gone over entirely to powered boats and is able to revive a moderate coal trade for a time. A novelty is a pair of boats crewed by women. These two travel the length of the canal five or six times carrying spelter to a new chemical works; the crew moor overnight at the old wharf and walk up to the Red Lion for a bath and a solid meal. At first hopes rise high in the hearts of the local lads but an attempt to board the butty one night is greeted by a well-aimed shove from a mop and the threat of a wallop from a windlass, and is never repeated.

Nationalisation of the waterways follows the end of the war. For a time it seems to make little difference; the same people stay in the same jobs and the same sort of trade continues, although many of the boats are now in blue-and-yellow livery instead of the green, red and black of previous years. The canal is omitted from any improvement schemes as it is not sufficiently busy to warrant money being spent on it. There are, however, some new users; the boys of Pendle School now have a canoe club and use a shed on the old wharf to house their craft, and about 3 miles (5km) of the water have been leased by an angling society from the Midlands whose members come down by coach on weekends in the season. More significantly, as it later turns out, two pleasure cruisers have appeared; a motor boat belonging to Rossvale's younger son and another to a couple who have bought one of the Victorian villas near the village.

The canal was not on Mr L. T. C. Rolt's itinerary and so escaped mention in his popular book *Narrow Boat*, but photographs of it appear in *The Canals of England*. The Inland Waterways Association mounts a local campaign to have the canal properly dredged and maintained which attracts the attention of the local press for a week or two. A chapter on the canal's history is published in Charles Hadfield's *The Canals of the Midshires*, the second volume in his 'Canals of the British Isles' series; but the canal employees and the villagers are mostly unaware of this moment of national fame. Nevertheless, these are among the events which ensure that the British Waterways Board, who now own the canal, will not downgrade it further, apart from classifying it as a 'remainder' waterway though with prospects of promotion to 'cruiseway' if this can be justified. So when a book called *Lost Canals of England & Wales* is published in 1972 this canal is not included in the contents.

If Brindley, Blake, Morgan or William Huckle were to revisit the canal in the 1950s or '60s they would find most of the scene fairly familiar. But a couple of decades later – in the mid-1980s, say – they might well feel almost entirely lost. The cut itself is still there and the locks are where they always were, but almost everything else has changed.

So, on a summer's day, what do we find? Firstly, a canal with plenty of boats. Several are clustered around Mill Lock, bright pleasure cruisers in narrow-boat style with cabins almost the length of the hull, and shorter white-hulled motor launches with windscreens and plenty of top hamper. More boats are moored on the site of the old wharf, now converted into the basin of a boatyard owned by Midshire Hireboats (proprietor T. J. Bridson), so a large painted board proclaims. The wharf buildings have been replaced by blue-and-white painted sheds and workshops, and the wharfinger's house has vanished. So, too, has Mill Lock cottage; there is no lock-keeper these days and the cottage was condemned because of its primitive sanitation. At first the lock itself seems little changed, but a closer look reveals that the balance beams, once huge baulks of timber polished by the friction of thousands of pairs of trousers, are now made of steel of half the dimensions, and the old rack-and-pinion paddle gear has been replaced by hydraulic machinery enclosed in steel domes. The towing-path has been widened and given a gravelled surface, though above the lock it reverts to a narrow grassy track.

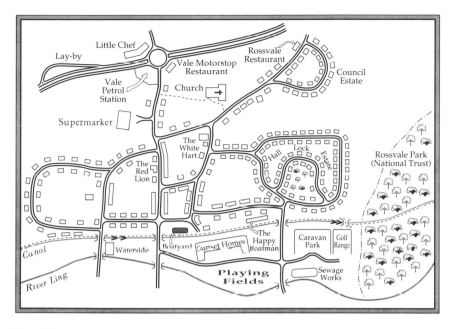

The village in the mid-1980s

Alongside the canal several new buildings have appeared, although not all of them are as new as they look. Here is a block of flats and maisonettes, five storeys high, with bow windows and little gables and a stretch of neatly mown grass separating it from the towing-path. This is Waterside; but if you hacked away some of the facing brickwork or examined the foundations you would find the vestiges of Smerdons' Mill. 'The Happy Boatman' is a new public house with a canalside terrace, its Butty Bar and Longboat Restaurant smartly done out with a décor of roses and castles; however, excavations in the adjoining car park would reveal the foundations of the Ship & Anchor. On the site of Bridson's private wharf there is now a large rambling house with tastefully laid out garden and a private mooring on the canal, the home of Mr T. J. Bridson, at present on holiday in Portugal. The boatyard and the haulage undertaking in the village will both run very effectively without him, managed by Kevin Rodd, a great-grandson of the Jason Rodd who once employed Joe Bridson as wharfinger's assistant.

Other new canalside buildings in the vicinity include a small group of single-storey retirement homes on the far side of the canal, the gardens securely fenced; and on the towing-path side the gap between the boatyard and the Happy Boatman has been filled by two bungalows, built by Bridson and sold for a small fortune. In a corner of the garden of one is a brick hut with concrete surround, much visited by boat crews as it is the only sanitary pump-out station on the canal.

The largest canalside development, however, is midway between Mill and Hall locks on the towing-path side: a recently built housing estate, Hall Lock Estate, sited on fields once watered by William Huckle, occasionally subject to inexplicable surface flooding. Despite this the houses are much in demand, especially by young executives working in the eastern terminal town – nicknamed 'Computer City' – where information technology has taken a firm hold. A new road links the estate with the village, a new concrete bridge taking the road over the canal and away to the south. Some years ago a caravan park was established opposite the estate fields and between the canal and river. However, many of the Hall Lock residents object to this park and are lobbying the local authority to have it removed. Not only do the caravanners' children swim in the canal by day, shouting and swearing, but at night the caravanners themselves dump their surplus rubbish in its waters or relieve themselves into it on their way home from the pub.

The village itself has seen great changes, and some are at least in part the result of the new-found popularity of the canal. The housing estate to the north and the supermarket in the High Street would have arrived canal or no, but why otherwise would there be a chandlery next to the newsagent's specialising not in pemmican and hard tack but in canalling leisure wear? Two village shops which lost out to the supermarket are now restaurants; one specialises in Mexican cookery and has no canal connection, but the other, Telford's Tuckaway, entices hungry holidaymakers with dishes such as Grand Union Duck Terrine and Gongoozle Pie – its tiny bar is of course called Worcester Bar, where a cocktail is served called Anderton Lift.

The Red Lion, however, has – metaphorically speaking – turned its back on the canal. Seeking to satisfy the wealthier parents of Pendle School and the rising executives of Hall Lock Estate, the management found the incursions of scruffily dressed canal boaters did nothing to improve the tone. They were therefore discouraged by a series of prohibitions; no children in here, no plimsolls or track shoes in there. If any penetrated the dining-room they were given the table by the serving door and the dishes they ordered were usually off. The Red Lion is now avoided by canal users; if the Happy Boatman and the Tuckaway do not appeal there is the Rossvale Motor Restaurant at the north end of the village or the Little Chef by the main road.

The village lost its railway service many years ago; the north–south main road has been widened and the canal bridge rebuilt in concrete and brick. A new east–west dual carriageway sweeps by a mile to the north, roughly parallel to the line of the old railway. There is a large roundabout at the junction of these two routes, where the Vale Motorstop and the Little Chef are situated. It is the improved road network that led to the building of Hall Lock Estate and encouraged Bridson to open his hiring boatyard.

Yet despite recent improvements the village retains some of its old-world attraction. Not all the old cottages have gone and the survivors are mostly in excellent condition, renovated and replastered, all now with garages and of course television aerials, and some with tactful extensions at the rear. On the roofs of a few the slates have been replaced by thatch, an intriguing reversal. Primrose Cottage in Mill Lane has been converted into an antique shop and a large Measham teapot may be seen in the bow window – it used to grace the parlour of the lock cottage, having been bought by one of the Tranters from a boatwoman for 10s (50p). Purchased from the last lock-keeper for £12 its role is to attract customers into the shop. If you ask, you will be given a mythical history and told it is not for sale; but if you offer £200 you might be lucky! A few years ago you could have bought ribbon plates and items of boat cabin furniture, including painted cans and stools, but now the supply of genuine articles has dried up and only poor modern replicas may be found.

The canal, directly or indirectly, is still one of the main sources of local employment. Two British Waterways Board workers live on the council estate: Bill Goodwin, lengthsman responsible for 8 miles (13km) of canal, and Mervyn Edwards, a crew member of the dredger on this section. The old lengthsman's cottage by the towing-path was demolished by Braygrove Breweries when the Happy Boatman was built. Goodwin patrols his length partly by car, driving from bridge to bridge and walking in between. The boatyard employs six people in addition to Kevin Rodd: four men repairing and maintaining the boats, and two women – Lucy Rowe in the office and Madge Faulkner who runs the boatyard shop selling books, souvenirs, confectionery, and also painted ware of startling ineptitude. Two part-time employees are taken on in the summer to help with the extra work at weekends.

Were it not for the canal, the staff of the Happy Boatman would not be where they are, and nor would the two women be employed by the village chandlery. But there is a larger group, most of whom live in the village, whose place of work is some 6 miles (10km) away, who also owe their living to the canal, even if indirectly: the staff of the South Midshire Waterpark. The waterpark is a recently developed leisure facility centred on Masser's Lake, one of the original reservoirs constructed on the advice of Robert Whitworth to supplement the canal's supply of water. Now it is used for sailing, water-skiing and angling, and has a handsome clubhouse with a children's playground and shop, an area reserved for birdwatching, a café and ample parking and toilet facilities. Including the twelve employees in the village, the canal is responsible in various ways for the living of about thirty people in all, a number only exceeded locally by the staff of Pendle School.

However, very few of the villagers actually use the canal themselves. Bridson uses his boat on occasions, usually to impress prospective suppliers or influential customers; Dr Baldwin of the Health Centre keeps a 35ft (10m) cruiser in the boatyard, and two of the

Local lads fishing

The 'ornamental lake', weedy and unkempt

Waterside residents own little powered runabouts. Pendle School has given up canoeing and concentrates entirely on ball games. Some of the local lads fish in the canal, now deserted by the angling society because of increased fees, but this is discouraged by Bridson because it hinders the smooth passage of boats. Bridson's hirers, and hirers from the other boatyards on the canal and its connecting waterways, mostly come from the London area or the North. The local people who seem to know the canal best are the members of rambling clubs who walk lengths of the towing path usually during spring and autumn. In past years the village schoolchildren used the towing-path for nature study, but the school was closed years ago and the children are now bussed elsewhere.

One important element of the local canal scene is Rossvale Hall. The present holder of the title sold the Hall and park to the National Trust fifteen years ago and went to live on the Algarve. Thus ended a long association with the canal during which the Rossvales had brought it near to bankruptcy and also saved it from extinction. The canal, of course, belongs not to the Trust but to the British Waterways Board. A previous Lord Rossvale had insisted that the Board erect fencing along the cutting so that access to the Park from the towing-path was impossible, and there is now disagreement between the Trust and the Board as to who is responsible for keeping the fencing in repair. Sections of the fencing have collapsed in recent gales so that here and there walkers and crew members can find their way through and explore the grounds, admire the peacocks and enjoy a free poundsworth of stately garden. The 'ornamental lake', known as Rossvale Wide, is of no interest to the Board, which is trying to persuade the Trust to take over at least some part of it. In the meantime it is weedy and unkempt, and any boat drifting off the navigable channel gets stuck in the mud very quickly.

How is it that this country canal has survived when so many others have succumbed and been abandoned? Many of them were closed when commercial traffic finished, and were dewatered and levelled so they could not take advantage of the upsurge of pleasure trade in the mid-twentieth century. But most of these were isolated from the main network – for example the Hereford & Gloucester Canal, closed in 1881, and the many canals of the West Country almost all of which were abandoned long before 1900. On some waterways, major engineering works such as long tunnels or inclined planes failed, and this accelerated closure as trade fell away. The canals had outlived their economic and commercial usefulness, and that was that. Who could have foreseen the 'leisure revolution' of the post-war years, or imagined that waterways such as the Wey & Arun Junction, abandoned in 1868, or the Basingstoke Canal, disused by commercial traffic since 1901, would be the subjects of ambitious restoration schemes today?

Soundly constructed and with no tunnels or costly engineering features, this canal had some initial advantages despite being restricted to narrow boats. Its link at the western end with the main network went far to preserving it, enabling some commercial trade to continue well into the twentieth century. Also it had friends at court, and other friends quick to see the opportunities for leisure and pleasure; the Inland Waterways Association supported it at a critical time as well. So like a few other rural waterways – the Brecknock & Abergavenny and the Southern Oxford canals for instance – it remained open, and its future seems assured.

But let us look for a moment more closely at what this future might be. It is unlikely that the British Waterways Board will endure in its present state for very much longer; with the reconstruction of the water industry it will become an anomaly, and mutterings of privatisation will be heard in the corridors. Peer, then, into the crystal ball. . .

. . . and here is our canal, now the property of South Midshire Leisure Heritage PLC. A brochure, obtainable from the company's smart suite of offices on the canalside, tells us that several miles of canal and canal bank to the east are ripe for development and proposals are invited from interested parties. In the vicinity of Mill and Hall locks, however, the company has undertaken the development itself, in line with its name and declared policy of recreating the best of our English heritage for the enjoyment and benefit of all. You may dine in the Old Mill, reconstructed in modern materials 400yd (365m) from the original site of the mill, and can watch the mighty millwheel revolving while you eat, encased in a perspex box and dribbling water pointlessly from its plastic paddles.

You may swim, play squash, pump iron or enjoy a sauna or sunray treatment in the James Brindley Sport and Leisure Hall, built on the old caravan site opposite the Hall Lock Estate. You may still hire a boat from the marina, electrically powered, centrally heated and colour televisioned with satellite dish mounted on the cabin roof. Or on weekends you can take a trip in the Midshire Traditional Horse-drawn Barge, steered by Judd Rodd wearing traditional boatman's garb of plush waistcoat, blue jeans, cloth cap and black trackshoes. If you wish to use your own boat on the canal, however, you have to make prior application to the South Midshire offices. You will need to purchase a canalpass (payment by credit card preferred) and will be issued, against a deposit, with a lock key. This resembles an ignition key, and without it you cannot operate the computerised mechanism that controls the opening of the lock gates. . .

The crystal ball clouds over. Wait a few seconds until it clears again, and then let us take one more look ahead. Peer more closely this time, into the further reaches. . .

The canal is still there, a narrow strip of water, clearer, less murky now. Here are the locks and the bridges, and there seem to be plenty of boats, mostly full- or half-length narrow boats, moving quietly along. Some of them are carrying freight packed in sacks or boxes, though none of them carries coal.

The canalside has greatly changed. Most of the taller buildings have gone, some swept away completely while others – including the

Sport and Leisure Hall – look unkempt and derelict. In place of the Waterside complex of flats and maisonettes there are now groups of small single- and double-storey houses, each with its own little patch of garden front and rear. The offices of the South Midshire Leisure Heritage Company have disappeared, and so has the company itself. Taken over by a so-called investment trust it was stripped of its various assets – the Old Mill restaurant, the Sport and Leisure Hall, marina and so on – which were sold off piecemeal, and was eventually liquidated.

The assets were unable to survive in what was now a contracting economy, and one by one they also failed. The site of the Old Mill was taken over by a local market gardener, and mushrooms are cultivated in what is left of the building. After many years of neglect the marina was bought by Gordon Bridson, not for sentimental reasons alone but because he could see that, as the oil supply beneath the oceans and in the Near East was exhausted, there would be a real future for electrically powered boats. No-one is interested in the Sport and Leisure Hall, where the maintenance and running costs always exceeded the income.

The Happy Boatman was also once a part of the South Midshire empire, and was eventually purchased by Garth Rodd, grandson of Judd. With the help of several villagers he demolished the old building and constructed in its stead a close replica of the Boat Inn at Stoke Bruerne; he had found a very old photograph of the Boat, taken in the days before the restaurant extension was added. At the rear he built a large stable block and a blacksmith's shop. After toying with various new names he decided that the Boat Revived was the most appropriate. . .

However, although nearly all seems flourishing and prosperous, there is a threat to the future of the canal of a kind that in this area at least has not been seen before. A few yards above Hall Lock a botanist on the staff of the Countryside Conservation Commission has discovered specimens of a notably rare water weed. Among these specimens he found a unique variant which, with a keen sense of the historical, he has named *Brindleiensis rubellina*. The members of the Commission are so impressed by the importance of the discovery that they are applying to the Courts for an order to have 200yd (183m) of the canal closed to navigation and all other activities so that the future of this weed can be assured. The argument that this measure would effectively cut off the village and the miles of canal beyond, and before long would lead to the complete closure of the waterway, has so far had no effect. The CCC has its priorities and strong government backing.

The best hope for the survival of the canal as a navigation seems to lie with Garth Rodd, or rather with his son Jim, a member of a sub-aqua club; Jim visits the canal most evenings with a knife, trowel, torch and sub-aqua gear, in an attempt to remove all traces of the weed before the Commission's next inspection. Whether he will be successful before the Commission obtains a temporary closure order no-one can tell. . .

The Coates round house on the Thames & Severn Canal

6 A Fenland Voyage

The Middle Level comprises the network of rivers and navigable drains between the Nene and the Great Ouse, and cruising these waterways can be a lonely occupation. You will see few boats, if any, on the move and on many stretches the high flood banks cut you off from the surrounding countryside. At times there is nothing to see except the sky above and the long straight ribbon of water ahead, with perhaps only a bridge in the distance to break the monotony. What are you doing here? Should you be here at all? Perhaps unknowingly you have passed a notice declaring this waterway closed to navigation, and at the end of this straight stretch there will be an impassable barrier or a waterfall at the edge of the world. . .

Despite the drainage works of Vermuyden and his successors, and the agricultural mechanisation of recent years, the Cambridgeshire Fens still maintain an air of mystery and are slow to yield their secrets. Here you are now, cruising on the Old River Nene, but it has no outlet to the sea, nor does it appear to have a source. At one end of it is Bevill's Leam and 'No through navigation'; at the other, Outwell Church, and Well Creek going off at a right angle. Moreover much of this Old Nene clearly does not follow a natural course as it tracks its way through the flat countryside in a succession of straight lines, curves and sharp angles. It is much more like a canal, except that the towing-path – or haling-way – is well above water level, sometimes on the top of the flood bank.

Anyway, here you are, exploring the Old River Nene. You entered the Middle Level system from the River Nene (the 'New Nene' presumably) below Peterborough, passing through Stanground Sluice on to the King's Dike. You negotiated the sharp bend at Whittlesey and continued along Whittlesey Dike to a sort of T-junction at Flood's Ferry. Here you decided (time being no object), to divert towards Ramsey, so you turned south. Soon there is a road on the left-hand side with an occasional car, farm lorry or tractor. The river swings south-west; ahead is a road bridge, and by it what was once obviously a wharf. Now there are trees and houses and another bridge. Moor here and climb the bank: this is Benwick.

The bridge is closed to vehicles so a visitor can stand here

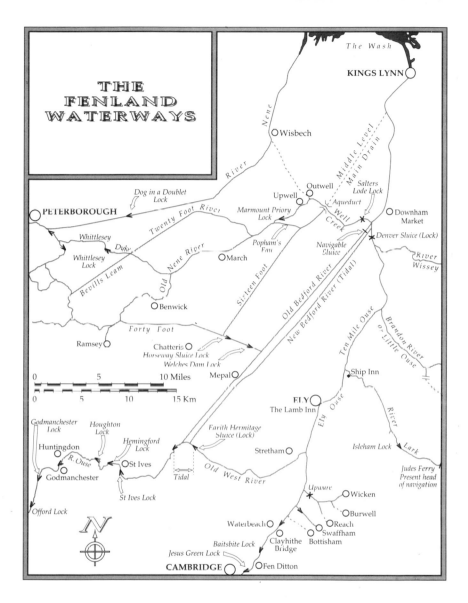

THE
FENLAND
WATERWAYS

The Wash

KINGS LYNN○

River _Nene_

○ Wisbech

○ Outwell   Salters
Upwell ○   Lode Lock
Middle Level Main Drain
Aqueduct
Well Creek
○ Downham Market
Denver Sluice (Lock) ✳

_Dog in a Doublet
Lock_

○ PETERBOROUGH

Twenty Foot River

Marmount Priory
Lock

Whittlesey
Dyke
Whittlesey
Lock

Bevills Leam

Old Nene River

_Popham's
Fen_
○ March

Sixteen Foot

Navigable
Sluice

_River
Wissey_

○ Benwick

Forty Foot

Ramsey ○

Chatteris ○
Horseway Sluice Lock
Welches Dam Lock

Old Bedford River (Tidal)
New Bedford River
Ten Mile Ouse
_Brandon River
or Little Ouse_

0   5   10 Miles
0   5   10   15 Km

Mepal ○

Ship Inn ●

ELY ○
The Lamb Inn

Ely Ouse

River Lark

_Godmanchester
Lock_
Houghton
Lock

Huntingdon ○
R. Ouse
○ Godmanchester

Hemingford
Lock
○ St Ives

_Earith Hermitage
Sluice (Lock)_

Stretham ○

Islcham Lock

Judes Ferry
Present head
of navigation

Old West River
Tidal

St Ives Lock

Offord Lock

Upware ✳
○ Wicken

○ Burwell
○ Reach
Swaffham
Bottisham

Waterbeach ○
○ Clayhithe
Bridge

Baitsbite Lock
Jesus Green Lock
CAMBRIDGE ○   ○ Fen Ditton

N

undisturbed and take in the scene. Beneath, the Old Nene makes a right-angle turn to head north-westward out of the village; in front, the wide High Street stretches southwards, and behind is a churchyard without a church but with gravestones tumbling in all directions. Along the High Street some of the houses are leaning backwards, one in particular at an alarming angle; there is one pub, the Five Alls, a couple of shops, few people. But why such

a wide street in a village on a route to nowhere, and where
is the church?

Some of the answers are not hard to find. A small plaque in
the churchyard reveals that the Church of St Mary the Virgin was
demolished in 1985, and a walk round the churchyard will soon
show you why. The soil here is peat, as it is throughout much
of the Fens. The foundation stone for St Mary's was laid in 1850,
and the architect was S. S. Teulon, well known for specialising in
ecclesiastical buildings; at the time the site seemed stable enough,
and indeed there had been a chapel there for two centuries. But with
improved drainage and new agricultural methods the peat had begun
to shrink and waste away; a few miles away are the Holme Fen Posts
which indicate how the surface level of the land has fallen 13ft (4m)
in about a hundred years.

St Mary's was built with dark brown stone from North Norfolk,
and had a nave, chancel and a stubby tower topped by a spire added
in 1902 when the whole building was restored. Real trouble became
evident in the 1950s. A. K. Astbury visited the church while working
on *The Black Fens* and wrote:

The church and children of Benwick in the early twentieth century (Cambridgeshire
Collection)

Inside the floor seems almost draped over the heaving earth; the pillars are out of plumb and there is a noticeable distortion in the windows. Outside it is clear that the tower is leaning away from the main body of the building, and even in the churchyard the headstones of the graves lean in all directions.

In 1967 the spire and part of the tower were demolished. The lurching and crumbling continued until in 1979 the church insurers insisted that only the chancel should be used. A full repair and restoration was estimated at about £100,000 – this for a church originally built to accommodate four hundred people but now with an average congregation of fourteen and an average collection of £3. A visitor at this time described it thus:

Looking west one experiences a remarkable sense of unreality as the pews slope towards each other, forming, as it were, a channel along the nave, while the entire west wall leans well to the right. There is a substantial crack from window-frame to roof of the west wall of the vestry (the base of what was once the tower). . . The builder repairing the chancel-door steps has found that the footings, of brick, are now above ground level due to the shrinkage of the peat.

The last service in St Mary's was held at the end of December 1980, with a congregation of thirteen. The church was then declared redundant and remained unused until it was demolished five years later.

There are very few settlements built on the fenland black peat; indeed, Benwick is the only one large enough to be termed a village. But Benwick's is a unique situation and the peculiarity of its very wide High Street should be explained. For, centuries ago, this High Street was a river, the West Water, a natural river that wandered across the fenland between the Great Ouse at Earith and the Nene. Then the construction of the two Bedford rivers by Vermuyden in the mid-seventeenth century effectively destroyed the West Water; between Benwick and March its channel is now occupied by the Old River Nene.

The present houses of Benwick High Street were built on the levées, or raised banks, of the extinct West Water. Sometimes however, according to A. K. Astbury, 'wedges of peat work themselves into the outer edge of a levée on which houses are later built.' This is why some of the houses slope backwards; doors and windows have to be re-hung and doorsteps and window-sills extended inwards. In 1950 the roof of the Old Rectory slid off, and about ten years later the Baptist chapel had to be demolished before it collapsed entirely into the peat.

Benwick is an old village. In 1221 it was recorded as having fifteen tenants, which had increased to thirty-two some thirty years later – being on a water trading route between Earith and Peterborough the village would have been relatively prosperous, at least from time to time. Benwick was described favourably by Isaac Casaubon on his tour of England in 1611, when he found the houses 'to the number of 200, all surrounded, as islands, and their inhabitants occupied in fishing and fowling.' Then in 1774, when the West Water was no more, Lord Orford sailed his fleet of nine small craft from March through Benwick on his voyage round the Fens. He described the village as:

> prettily situated on each side of the river, the grounds having a fertile appearance. In passing through a bridge in this village, which necessarily occasioned some delay, the people assembled as usual to see us pass. A number of children, crossing together near a neighbouring cottage (a school), added to the simplicity of the scene, which had much the appearance of some of the best Flemish landscapes.

Today this impression cannot be recaptured as the cottages on the north side of the Old Nene have gone, and indeed there are no really old buildings in the village. Possibly at that time Benwick resembled the villages of Upwell, Outwell or Nordelph where houses face each other across a river, as they do in parts of Holland.

Until recent years Benwick depended on the Old Nene for its supply of fuel. Strings of barges brought turf, dug in Ramsey Fen, to a wharf near the church. Here it was unloaded and stacked in ricks to be carted round Benwick and nearby villages by merchants who sold it in hundreds. Few of the cottages had separate bread ovens; the custom was to burn the turf in the oven until it was red hot, then sweep it out and put in bread, pies and milk puddings and cook them all together. The river was effectively the main transport route to and from the village until the arrival of a Great Eastern Railway branch line in 1898. The station – goods only – was built beside the river by the road bridge and became a busy rail–water interchange depot. Here coal and timber were offloaded into lighters to supply the pumping engines in the Fens, and potatoes and other agricultural produce were brought for rail delivery to connections at Peterborough and March. The railway was closed under the Beeching cuts in 1964.

Benwick has lost six of its seven pubs, including the largest of them, the Boot and Slipper, which stood near the river until it was demolished in 1971. However, it is by no means a dying village. It has none of the half-timbered cottages, antique shops and smart new estates found elsewhere but its workmanlike, even dour, exterior hides a wealth of fenland history. Nowhere else will you find a village like this.

Now continue westward on the Old Nene towards Ramsey. The river swings south-west, and for the next 2 miles (3km) or so you are navigating through what was once Ramsey Mere, a large and beautiful lake famous for the number and size of its pike. 'Here Pikes of a wonderful bigness are caught and tho' it is perpetually haunted by Fishers, Fowlers and Pochers, yet there's an inexhaustible store of Game left,' wrote Emmanuel Bowen in 1756. Ramsey Mere, like the even larger mere near Whittlesey, was drained in the mid-nineteenth century and its bed is now rich farmland.

At the southern edge of the mere the Old Nene makes a sharp bend. By Wells' Bridge, Vermuyden's Forty Foot Drain heads off due east in a straight line; less than a mile along is the settlement of Ramsey Forty Foot, once a little inland port and with a couple of good eighteenth-century houses surviving. If you carry on along the Old Nene, however, another waterway branches off to the south. This is the High Lode, which in about 1½ miles (2.4km) leads to the very edge of the town of Ramsey – and then suddenly vanishes. There are new moorings here by a tall building which was once a mill but is now converted into private dwellings – tie up and look more closely at the brick arches through which the water of the lode disappears. It is then easy to walk into the town.

The main street of Ramsey is called the Great Whyte, a name recorded in the thirteenth century. It is exceptionally wide; indeed for almost all of its length it consists of two parallel roadways with islands along the middle. Until 1852 it carried the waters of a stream known as the Bury Brook, crossed by the Great Bridge with a single pointed arch. Between 1852 and 1854 the stream was roofed over – but it is there still, flowing beneath the traffic, and emerges on the south side of Ramsey. The Bury Brook is a natural stream, rising near the village of Wistow and flowing past Bury, through Ramsey and into the High Lode. Before the lode was constructed the brook continued, to join the Old Nene about 1½ miles (2.4km) to the north-west. The lode replaced this stretch of it as a navigation and so the brook north of Ramsey disappeared.

Ramsey in earlier centuries was an inland port. It is likely that stone for Ramsey Abbey, once one of the richest establishments in England, finished its journey via the Bury Brook, and water transport would have served both the market and the fair. In the Ramsey Rural Museum there are reproductions of paintings showing the Great Whyte with the stream running through it and several small boats and boathouses. However, there are few really old houses along the Great Whyte as a fire in 1731 destroyed almost all the western side, eighty houses, shops, barns and granaries being burnt to the ground. The abbey was demolished in the sixteenth century and a school now occupies the site; St Thomas's Church was originally the abbey guesthouse and contains much thirteenth-century work. Across the green is the fifteenth-century gatehouse, owned now by the National Trust. The Rural Museum is at the end of a lane opposite the cemetery on the road to Ramsey Forty Foot.

Now return to your boat at the bottom of the High Lode and turn it – no problem now that the navigation here has been widened. At the junction, turn left and make for Nightingale's Corner and the southern edge of Holme Fen. Follow the river northwards; until the mid-nineteenth century this strange countryside of flat fields and straight roads was under the water of Whittlesey Mere, then England's largest lake.

Some years after the mere was drained in 1851, the parish of Holme was created, 4,900 acres (1,983ha) of fen. Much of the parish was subject to flooding from time to time and the tracks which served as roads were often impassable. When the Rev Horatio George Broke came to Holme in 1895 he found that many of his parishioners lived more than 3 miles (5km) from the church and in the winter months it was impossible, or at least very difficult, for them to come to services. He reasoned that if they could not come to church then the church had to go to them, as it did in the parish of Stretham where the vicar had purchased three vans and six horses, at his own expense, to put the church on the road to his outlying parishioners. But the Holme roads were too rough and wet for this. What Holme did have, however, was 9 miles (14km) of navigable waterway. Broke was a resourceful and determined cleric and ordered a specially

Ramsey. The head of navigation

Upwell on the River Nene

built flat-bottomed barge with a large room, 30ft long, 9ft wide and 7ft high ($9 \times 3 \times 2$m) on top of it. For this, William Starling of Stanground charged the diocese £70. It was completed by April 1897, and dedicated by the Bishop of Huntingdon on 5 April as St Withburga's Church – Withburga was a sister of Etheldreda, patron saint of Ely Cathedral. The floating church was equipped with altar, lectern, small stone font, prayer desk and a miniature harmonium, and had room for a congregation of nearly fifty. It was licensed for baptisms only, and in later years possession of a baptismal certificate from St Withburga's or the Floating Church was greatly envied.

Three stations in the parish were used in turn for the holding of services. The church was moved by a horse called Boxer, owned by its caretaker Mr Langley who was also landlord of the Exhibition Inn at Stokes Bridge, one of the stations. Sometimes a Mr C. Sandby towed the church for short distances. Early on Sunday mornings the vicar rode out, on horse or bicycle, to where the church was expected to be. Sometimes if the weather or river conditions were bad it wasn't there; sometimes it was, but the congregation was missing.

The church floated in the parish for over seven years, during which time seventy infants were baptised on board. For a time there was a floating choir and also a bible class, though neither lasted for very long. Hard winters and dry summers, when there was insufficient water for the church to move, discouraged the congregation, and the Rev Broke's health was also deteriorating. Then in 1904 the vicar of Manea asked if he could borrow the church for use at Welches Dam, a settlement near the point where the Forty Foot Drain meets the Old Bedford River. St Withburga's remained at Welches Dam for two years and three more baptisms; then it was returned to Holme, in dilapidated condition. However, it was no longer needed or usable; the furniture was removed and it was towed back to Starling's yard.

But this was not the end of the story. In 1907 it was bought by a group of young men who converted it into a houseboat, renamed it 'Saint's Rest' and moored it at Orton. Here it provided a happy holiday resort until in 1912 it sank near Orton Staunch a few miles from Peterborough.

Now you may return to Nightingale's Corner, head east along the Old Nene, past Benwick, past the junction with Whittlesey Dike at Flood's Ferry, beneath the bridge at Botany Bay and into the town of March. Two market places – one of them now a car park – developed alongside the river, the south side market in the sixteenth century and

Outwell on Well Creek

the north side somewhat later. Apart from St Wendreda's church with its magnificent angel roof which lies a mile (1.6km) to the south, the best of March is probably seen from the river. However, in the last century or so the town's growth has been mainly along the north–south axis, while the river crosses from west to east.

Carry on north-east through March, past the junction with the Twenty Foot and the junction with Popham's Eau, a cut made to the order of Lord Chief Justice Popham in 1605. Soon you come to Marmont Priory lock and the southern fringe of the village of Upwell.

Since the waterway that once flowed through the centre of both Benwick and Ramsey has disappeared, to navigate through Upwell and its companion village of Outwell is a treat indeed. For the Old Nene flows through the length of both villages, with a road on either side of it and the houses and cottages – many of them handsome buildings from the eighteenth century – facing each other across the roadways and the river, which here forms the boundary between Cambridgeshire and Norfolk. St Peter's overlooks the Nene about half-way along, a fine old church with a roof of angels. Then at the end of a 4-mile (6.4km) stretch through the villages you arrive at Outwell basin where the Old Nene ends. However, also flowing into this basin, in the shadow of St Clement's church, is the Well Creek, a fascinating little waterway which after about 5 miles (8km) leads to Salter's Lode lock – and thence to the Great Ouse, downstream to King's Lynn, upstream through Denver sluice to Ely, St Ives, Huntingdon and Bedford, or via the Cam to the Cambridgeshire lodes and to Cambridge itself.

Outwell Basin was an important waterway junction for several centuries, with the Old Nene, the Well Creek and the now sadly defunct Wisbech Canal all meeting here. In medieval times the Old Nene shared the channel hereabouts of the Wellstream (which gave its name to the villages of Outwell and Upwell). Stream and river wandered across the Fens from Littleport and eventually made an outfall close to Wisbech – the line later used for the Wisbech Canal. What was left of the Wellstream almost all disappeared as a result of the seventeenth-century drainage; by the time the Wisbech Canal was begun, the Wellstream was dry and houses had been built along its bed – these had to be demolished to make way for the canal. It was eventually opened in 1796 and though it ceased trading in the 1920s it was not filled in until some forty years later – now the canal lies under the gardens of the houses on the west side of the basin. Its line followed that of the older river and still marks the county boundary between Cambridgeshire and Norfolk in this area.

From 1883 onwards the canal's trade suffered from competition from the Wisbech & Upwell Tramway, a narrow gauge roadside railway for passengers and freight which ran alongside the canal. In its early years it carried about 3,000 passengers a week, and quite eclipsed Mr Whybrow's Packet, a horse-drawn passenger service on the canal which, like the railway, charged 2d (1p) a trip but was much slower. The railway endured until 1966.

Well Creek is a remnant of what Professor H. C. Darby described as 'a great water highway between Lynn and the midland counties'. As already mentioned, drainage operations all but extinguished the creek, and from time to time it was further threatened with closure in the interests of road widening schemes. Navigation was sometimes difficult if not impossible; one Christmas Eve in the 1960s Ted Appleyard, skipper of the miniature oil tanker *Shellfen* which supplied fuel to isolated pumping stations in the fenland, had to get out and push his vessel through the mud. Since then the Well Creek Trust has been formed to promote the maintenance and use of the waterway, and the creek is now open to what little navigation there is in the Middle Level.

So here you are in Outwell Basin, at the mooring stage by the one-time entrance to the Wisbech Canal, with the Old Nene and Well Creek heading away at right angles before you.

Lord Orford on his 1774 voyage described Outwell and Upwell as 'populous towns', though their inhabitants were subject to attacks of the ague, 'the reigning disorder in those parts'. He was struck by the number of old women and also by 'that disagreeable arrangement of features called ugliness' which afflicted the sex in Outwell, Upwell and March. However:

> On enquiry our surprise ceased when we were informed that this part of the country was settled by a colony of Dutch about the time of the revolution, which accounted for the squat shape, flat nose and round unmeaning face which prevails amongst the inhabitants.

Judge for yourself whether these comments still hold good today!

Leave Outwell Basin by the tree-lined Well Creek; you will soon cross the only aqueduct in the Middle Level which takes the creek across the unnavigable Middle Level Main Drain, constructed in 1848 to take surplus water from the Sixteen Foot to a point on the Great Ouse some 6 miles (10km) nearer the sea than hitherto. The water of Popham's Eau flows into Well Creek at the village of Nordelph, the one-time terminus of Mr Whybrow's Packet. The houses of Nordelph also face the waterway, which is accompanied by the A1122 from Outwell to Salter's Lode.

The lock at Salter's Lode is only usable for a certain period each side of high tide, and leads you into trickier waters. Within a few hundred yards the Old and New Bedford rivers debouch into the Great Ouse; away to the right is the complex of sluices at Denver, the keystone of the drainage of these fenlands. Enter the Great Ouse, turn upstream and make for the lock at Denver Sluice. Beyond the lock the waters of the river are non-tidal; this is the last stage of your exploration of the fenland's country canals. Oddly enough, the word 'canal' is hardly used in the Fens, except with reference to the defunct Wisbech Canal – 'river' 'drain' and 'cut' are the common terms. Yet if a canal is 'an artificial watercourse for navigation or irrigation', as the dictionary defines it, then canals are what you have been cruising along since entering the Middle Level at Stanground, except for a few stretches of natural river on parts of the Old Nene and Well Creek. Much of the course of the Great Ouse is now artificial too,

Upware Lock from Reach Lode in the 1930s (Cambridgeshire Collection)

identifiable by the long straight reaches, heavily embanked, which cut directly across the Fens where once the natural river wandered and shifted and spilled over into the fields.

As you head south, the Great Ouse tributaries enter from the east; first the Wissey, then the Little Ouse and, midway between Littleport and Ely, the River Lark. The Wissey follows a mainly natural course, but the Little Ouse and Lark in their lower and middle reaches are mostly artificial. All are navigable; the Wissey to Stoke Ferry, the Little Ouse to Brandon and the Lark to West Row. Less than a century ago the Little Ouse was a busy navigation to Thetford, and the Lark to Bury St Edmunds, but nowadays few boats penetrate these waterways; they provide quiet cruising, with occasionally a pleasant, tree-lined stretch, although at times the cross-winds present problems.

Superb views of the cathedral can be enjoyed from Ely's waterside; the moorings are good here, and there are plenty of places to eat. After Ely continue southwards, past the entry to Soham Lode, and take the left fork (as it were) into the River Cam at

Pope's Corner by the Fish and Duck. Now your exploration of the Cambridgeshire navigable lodes will soon begin; these are a series of artificial waterways, some of them first cut in Roman times, which link the villages of the fen edge to the Cam and thence to the Great Ouse and the sea at King's Lynn, or to the rivers and canals of the Midlands and the North.

The guillotine-gated lock at Upware gives access to three of the lodes: Reach, Burwell and Wicken. Here there used to be an old whitewashed pub, the 'Lord Nelson' – familiarly known as the 'Five Miles from Anywhere – No Hurry'. It was popular in the Victorian era with both Cambridge undergraduates and the lightermen of the Fens, and in 1851 it became the headquarters of the 'Upware Republic'. The Republic was a society founded by a number of Cambridge undergraduates who enjoyed fishing, shooting, drinking, smoking and talking, both with each other and with the countrymen and watermen who frequented the pub. The Republic had various officers, including a consul, president, State chaplain, Minister of Education, professor, interpreter, champion and tapster. The land-lord, Mr Appleby, was vice-consul.

The Visitors' Book records many names among the membership which later became famous, including the physicist James Clerk Maxwell; H. A. Morgan, later Master of Jesus College; Samuel Butler, the satirical novelist; and John Eldon Gorst, MP, Solicitor General. It also records catches of fish – 38lb (17kg) of pike; shooting triumphs – fifty linnets, four owls and many 'Snip' (or snipe); various eating and drinking feats; and the means employed to reach the inn – sailing, skating on the frozen rivers, or rowing in 'funnies' or 'the tiddledy-widdledy thing' as the locals described a single scull.

The Visitors' Book ends in 1856 and the Republic seems to have gone into abeyance. However, ten years later it was revived, but this time as a monarchy when Richard Ramsay Fielder, MA, of Jesus College, proclaimed himself King of Upware. 'A conspicuous figure in red waistcoat and corduroy breeches, drinking and fighting with the bargees as his strange humour led him', he held court at Upware for several years, the centre of a fluctuating group of undergraduates who drank gallons of his punch and talked and argued long into the night. Fielder had a high opinion of himself as a poet, and his works – subscribed 'His Majesty of Upware' – were printed on single sheets by J. H. Clements of Ely and distributed locally, especially around the public houses. They included such titles as *To the memory of Arthur, Duke of Wellington, Welcome the Daughter of the Czar, In commemoration of Etheldreda, Queen and Saint, The Stars – are they icicles or stalactites?* and *My Beautiful Cow.*

Fielder was no 'mute inglorious Milton' – he certainly wasn't mute, he took on plenty of glory and his verse is pretty dreadful – but he added to the excitement of life locally, being both argumentative and aggressive and reputedly never beaten in a fight. When the railway came, however, the river traffic diminished and fewer lightermen called at the old No Hurry; eventually Fielder left the district and moved to Folkestone where he is said to have become 'a reformed character' living in the odour of sanctity.

Upware was also popular with students of natural history, and the pub was a convenient base for botanists and entomologists from Cambridge and elsewhere, researching the rich natural life of the old Fens, especially at nearby Wicken. During the Great War the No Hurry was used to house German prisoners – many windows were broken, including some inscribed with the names of members of the old Republic; the thatched roof was replaced with corrugated iron. Later, the pub caught fire and was completely destroyed; it was some time before a replacement was built, and this would have been more at home on a suburban stretch of old Great West Road. Writing in

The entrance to Wicken Lode

1979, a few years before this perpetration, Professor E. N. Willmer commented elegiacally on what he saw:

> There is now no republic, no kingdom, no ferry and indeed no inn. . . Upware is no longer miles from anywhere. Hurry lies just round the corner, for motor cars speed along its narrow roads and motor-cruisers hustle each other into the narrow lode to ensure their mooring places. The subtle and undefinable river smell of drying mud and rotting vegetation drifts away before the urgency of diesel fumes and tarmac. The coots lurk more coyly and the tail-flicking moorhens dangle their trailing feet as they skim the water in their haste and alarm at the barking hooter or eructing horn of the oncoming cruiser. The leisurely chains of the ferry crossing have even rusted away into oblivion.

Only in the winter months is it possible to catch a hint of what Upware was like in its days of glory – if you stand with your back to the present pub, that is.

Once through Upware Lock you are in Reach Lode, with a line of boats moored alongside the banks, safely tucked away from the more mobile River Cam. Soon there is an opening on your left, the entrance to Wicken Lode.

Wicken Lode is only navigable by very light craft. It leads to Wicken Fen, a wetland nature reserve now owned by the National Trust and one of the very few surviving remnants of the ancient fenland. It does not survive in its 'natural state' however, but is the result of several centuries of cultivation and management. Today it is used mainly for the cultivation of sedge, litter and peat, as well as being protected as the habitat of an enormous variety of animal and insect life and a vast range of wetland plants. Drainage channels run through the fen, some of which were used in the past by small boats carrying away the peat and sedge, but Wicken Fen has never been subjected to systematic agricultural drainage. Consequently it stands several feet higher than Adventurers' Fen on the southern side of the lode. The drainage mill, or wind pump, at Wicken has been restored to working order.

Burwell Lode branches off Reach Lode at Pout Hall. It is just over 3 miles (5km) long and in its trading days could take vessels up to 50ft (15m) in length and 13ft 6in (4m) in beam, and dates from the seventeenth century when it replaced an older waterway on its southern side. The lode became particularly busy with trade when T. T. Ball's chemical works opened on its banks in the late 1850s, a mile below Burwell. Ball's works developed into Colchester & Ball's Patent Manure Works in the 1890s; they ran two steam tugs and two gangs of lighters. Their successors, Prentice's, also had a boat-building business in Burwell; they were building lighters until 1920 and repairing them until 1936. Phosphates were brought up from King's Lynn and fertiliser taken to Lynn for export or delivered around the fenland farms. Other cargoes included coal, stone and materials for brick-making, and bricks were carried on return from the Burwell Brick Company adjacent to the fertiliser works. The beet traffic continued until 1963, the last regular fenland water trade.

At first, Burwell Lode may not seem very attractive: an almost straight stretch of water, unprotected by trees and rather higher than the surrounding farmland, and featureless except for rows of pylons, a distant church tower or water tower. But then you come to Burwell, and a T-junction – to the left is 'The Weirs', an arm of the lode used mainly by lighters in the past; to the right 'Anchor Straits' where centuries ago coastal vessels would moor. Once, both these arms were lined with wharves and warehouses – now, almost all the wharves and hythes are filled in and only two warehouses remain. Recent development along the lodeside has obscured nearly all evidence of the trading past.

Burwell itself is a large village of the fen edge, served by its country canal for over 300 years. Be prepared for plenty of walking; the splendid church is over a mile away at the southern end of the village, and there are many other attractions as well. To start with, the Royal Commission on Historical Monuments lists 120 buildings in Burwell dating from the late sixteenth and seventeenth centuries, many of them houses in North Street. Usually substantially built of good quality materials, they were probably associated with the trade of the lode. Behind the church is the site of a castle hastily built by the twelfth-century King Stephen to counteract the rebellion of Earl Geoffrey de Mandeville. In August 1144 the Earl led an army to attack the castle, but was wounded by an arrow and died shortly afterwards. The rebellion subsided and the castle was abandoned.

On the west side of North Street is the little district of Newnham, a fifteenth-century rectangular planned settlement; and on the north side of Newnham is the Hythe, once Burwell's public wharf. At one time there were canal arms along either side of the Hythe, with a basin at the end of the southern arm. And elsewhere there were more than twenty basins or short canals, many of them connecting with the houses on the west side of North Street.

Many people find today's fenland unattractive, even repellent: vast flat fields of wheat, oats or sugar beet; only straggly bits of hedge, and maybe here and there just two or three trees; straight concrete roads, and long lines of ditches intersecting the landscape; they consider the wide, ever-changing skies to be the only saving grace. But was the countryside that this has replaced so very beautiful? It was valued by the wildfowler and the punt-gunner, but what was it really like? As you head along Burwell Lode, imagine the scene through the eyes of Alan Bloom who took over Priory Farm in Adventurers' Fen in 1939:

On the left of the lode was a sea of reeds, dead reeds, with their plumes tossing and waving in the wind. The outline of Reach Lode could just be seen to where it joined Burwell Lode. For almost a mile it was nothing but reeds except for clumps of bushes here and there. Among them, close by, there was the glint of water. A few half-rotted hay stacks stood near the bank – the fen litter, I supposed, for which there had been no sale these last twenty years.

On the right was Priory Farm, not very far away. I could see the entrance to another drove quite close. . . It was not black and muddy, but covered with dead grass. Its outline was quickly lost in bushes showing bare and gaunt, whose twigs rattled like bones in the wind, whose lower parts were hidden in the reeds and sedge of many years' growth. In places the bushes were thick and tall, and between them were patches of dense reeds. . . A little more to the right was a black mass of bushes. . . and between this and the farm buildings lay two arable fields.

In what seemed to be the centre of this desolation stood an old pumping windmill with its bare sails outspread, unmoving, I felt depressed. It seemed as if the bushes and reeds were moving nearer to the farm, nearer to the remaining cultivated land. The distant mill seemed to say 'I was put here to pump water so that farmers might reap what they sowed, and now look at me! The water beat me, and the bushes came, and they are both coming up there slowly but surely.'

However, three years later this desolation was already changing

Reach Lode from Reach Hythe

into rich food-bearing land, helped by the reclamation efforts of the War Agricultural Committee. And on one special Saturday a gault barge brought King George VI and Queen Elizabeth along Burwell Lode to inspect what had been done and give it their royal approval.

> Land girls from Manchester showed that they knew how to dig round bog oak and drive Fordsons. Stolid Fenmen down in the slub of a six-foot dyke grinned and tried hard to speak King's English when spoken to. . . The assembly saw that this forsaken land could be drained and farmed and made to grow good crops.

The unimaginable had come to pass: royalty had sailed on Burwell Lode and set foot on the black soil of the Fens – and for once, royalty was aware that Newmarket and Sandringham were not all that East Anglia had to offer.

Reach Lode itself is now little used, although a few boats do venture to the Hythe to celebrate the ancient festival of Reach Fair. The village of Reach sits at the end of the Devil's Dyke, a massive defence barrier 7½ miles (12km) long cut some time between 400 and 600AD to deter invaders from the South West. Originally the dyke extended to the head of the lode and settlements grew up on either side of it, but over the centuries the easternmost few hundred yards were cut away so that the two rows of houses now face each other across Fair Green.

Between the fourteenth and eighteenth centuries Reach was quite a busy little port, importing goods from the Continent via King's Lynn and exporting agricultural produce, peat and, in later years, coprolites. For thirty years or so the coprolite industry was very important in South Cambridgeshire and employed hundreds of labourers; coprolites in Cambridgeshire were not, however, the dictionary definition 'fossilised dung', but 'phosphatised nodules of clay, shells, sponges or other fossils which are found on top of the gault clay'. They were dug up, often from a depth of 20ft (6m), and processed into fertiliser; the industry lasted until the 1880s and has left its traces in trenches and mounds in fields around Cambridge.

Vic Jackson's gang of lighters in March, 1937. Work was held up by the water shortage

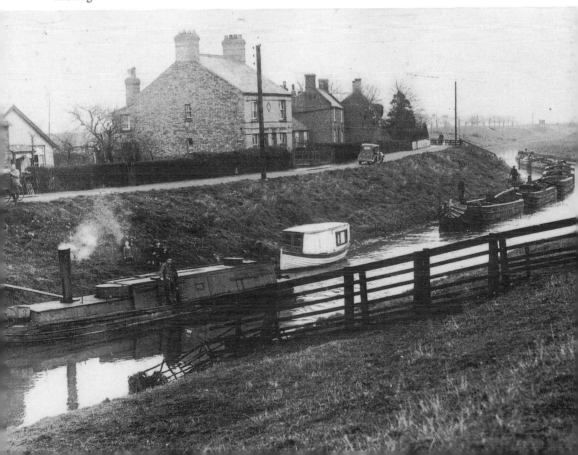

In the Middle Ages, Reach was important as one of the nearest inland ports to London. It was also vital for Ely Cathedral; stone from the quarries at Barnack was landed at Reach, and iron for Adam de Walsingham's Octagon was bought at Reach Fair. However, in the mid-nineteenth century development of trade at Burwell affected Reach adversely, and the opening of the Cambridge – Mildenhall railway in 1884 put an end to regular trading. Vic Jackson, the last of the great Fen lightermen, took the last cargo of clunch – a chalky building material quarried locally – from Reach to Peterborough in the 1930s; the voyage took three days.

Reach is a quiet little village with few old houses; it suffered frequently in its history from fires caused by the combination of thatched roofs and the use of peat as fuel. The Hythe is an outcrop of chalk 160 yards long and 36 yards wide (146 × 51m) jutting out into the lode. Looking out over the flat fields, with bright yellow rape in the foreground in the summer, it is not easy to visualise sea-going sailing vessels, or even gangs of lighters carrying 60 or 70 tons of coal, making their way up the narrow lode to tie up and unload to left or right, or manoeuvering into one of the many private basins nearby. Now the lode is narrow, shallow and weedy and even the small, shallow-draught motor cruisers have trouble enough scraping along to moor here on May Day.

The Hythe was constructed as a wharf to serve the busy trade of the port, quantities of chalk rubble being laid on top of the fen soil. Recent times have revealed the authorities' disregard for the rich history of the Hythe, however: they have dumped a sewage plant on it – the best you can do is let your imagination roam with your back to this; you might be on the edge of the world or, as the Rev Edward Conybeare, historian of Cambridgeshire, wrote in 1910: 'at the end of all things'.

Swaffham Bulbeck Lode has its entrance 2 miles (3km) further up the River Cam but is today unnavigable, apart from a stretch used as moorings. In the late seventeenth century, however, the little inland port known as Newnham developed at its head, and trade increased in the following decades, centering on the handsome Merchant's House; this became the headquarters of the firm of Bowyer and Barker, exporters of grain to the port of King's Lynn and importers of a variety of goods including coal, wine and oil. When Thomas Bowyer became the sole proprietor in 1805 he built warehouses, terraced cottages for his workers and a house for his agent. Newnham, now known as Commercial End – the commercial end of the village of Swaffham Bulbeck – was a busy little place until the railway opened nearby in the mid-nineteenth century. Trading on the lode continued into the 1870s, but profits were diminishing and the Merchant's House and other buildings were sold.

Commercial End is one of the most architecturally satisfying places in Cambridgeshire with its mixture of rustic cottages, workmen's dwellings, fine houses – including the Merchant's House – and a handsome maltings. No fewer than twenty-five of these are listed by the Royal Commission on Historical Monuments. One of them is the undercroft of a Benedictine Priory, founded by Isobel de Bolbec soon after 1200AD and now concealed within an eighteenth-century farmhouse – once it was the most important ecclesiastical establishment in the area.

The last of the Cambridgeshire lodes is Bottisham, its entrance a few yards below Bottisham Lock on the Cam. At the head of the lode, now unnavigable, is the little village of Lode where once there was a wharf and basin served by fen boats until about 1900. The lode is fed by Quy Water, which borders the grounds of Anglesey Abbey; the abbey has one of the most splendid gardens in the country, a magical blend of the formal and the picturesque, created in the mid-twentieth century by Lord Fairhaven, and to quote the Historical Monuments Commission, 'extensively furnished with statuary and urnage'. In the northernmost corner of the grounds is Lode Watermill, built in the early nineteenth century and restored to full working order. Both abbey and mill are now owned by the National Trust.

Bottisham Lode may date from the Roman occupation although, like the other lodes, it has been re-cut and sections re-aligned many times over the centuries. Until 1968 there was a fine example of a staunch or flash-lock, built in 1875, but sadly this is now no more, although the brick chamber may still be discerned.

Other Cambridgeshire villages, notably Cottenham and Swavesey, were also connected to main rivers by lodes and thus were served by water transport in medieval times and later. No county has so many old inland ports or so many little country canals, and even if they are not picturesque, they were certainly invaluable at the time to their local communities. The survivors are now comparatively little used by boats and from time to time their existence as navigations is threatened. Only continued use and local vigilance will ensure their future as navigable waterways, and prevent their decline into high-level drainage channels where neither fish nor ducks will find much joy.

7 Lost Country Canals

In 1845 there were fourteen canals totalling over 140 miles (225km) of navigation at work in the counties of Somerset, Devon and Cornwall, as well as a handful of short waterway connections. Two more canals, never completed as planned, were already out of use. Today just 33 miles (53km) remain and nearly all is restricted to small or unpowered craft. You can trail your boat to the Bridgwater & Taunton Canal, or enjoy a ride in a horse-drawn narrow boat along the surviving length of the Grand Western. When the tide permits you can enter the sea-lock at Bude or the tide-lock on the Exeter Ship Canal – and that's about all. Otherwise to visit the canals of the South West you need a car, good maps, stout shoes, plenty of time and enough imagination to recall to life the bits and pieces of long-abandoned waterway which are all that you will find.

As elsewhere, it was the lure of quick profits for a comparatively small outlay that inspired the creation of these canals. Often there was the prospect of coal carriage to market towns; also building materials could be transported cheaply to villages and farms; agricultural produce could be shifted easily to markets, and fertilisers could be distributed. There were grandiose schemes for linking the Bristol and the English channels, including one for a canal to take 200-ton vessels which would cost nearly £2 million to construct. At the other end of the scale were the little tub-boat canals designed to run along the high hillsides of Cornwall and Devon, carrying sea-sand in trains of 'boats' – rectangular boxes with wheels, linked together – to fertilise the barren upland fields. But only one of these schemes ever repaid its investors – the Somersetshire Coal Canal, the least countrified of all, which served the thriving Somersetshire coalfield for more than eighty years. Moreover it connected with the Kennet & Avon and hence with the inland navigation of England as a whole; and it was a conventional narrow boat canal as well.

The story of the South West canals shows that agricultural trade was never sufficient to keep any waterway alive; as soon as a railway appeared the canal trade withered and died. Despite impressive prospectuses the canal companies were mostly shoestring undertakings: local bankers and tradesmen might invest in them, but the

The Bude Canal near the foot of Marhamchurch Inclined Plane

heavyweights from London and the larger cities or the major canal companies were not interested. Hence standards of maintenance were generally low and money for repairs or the payment of wages was frequently unforthcoming. In March 1831 tolls on the Bude Canal totalled £41 8s 9d, while wage arrears amounted to £136.

Most of these canals were short-lived. The Grand Western Taunton to Lowdwells section lasted twenty-nine years; the Chard Canal, completed as planned, twenty-six years; and the Glastonbury Canal only twenty-one. The uncompleted St Columb Canal endured for four years at most, though this was still better than the Dorset & Somerset, of which 8 miles (13km) of a branch were constructed at a cost of some £66,000 but on which no vessel ever traded.

Nevertheless, the canal companies and their surveyors chose some enchanting stretches of countryside for their undertakings, and produced some extraordinary feats of engineering – not always successful, though doubtless spectacular in their time. They were often the work of James Green, an imaginative and innovative engineer whose plan for the Chard Canal, for example, was described by one of the promoters thus: 'a more ridiculous, ill-advised, inefficient, incompetent scheme was never devised'. But Green, who worked on the Exeter Ship Canal, the Grand Western, Torrington, Bude and Liskeard & Looe, did not merit such criticism. He knew very well how to solve the problems presented by the irregular terrain through which these canals were intended to run, and it was not his fault that funds to keep the works in good repair were so often lacking. Much of what he did is now almost obliterated, for example the great inclined planes of the Bude Canal, the boat lifts on the Grand Western – though today inclined planes and lifts function effectively on a scale he would never have dreamed of, on the continents of Europe and North America, if not in his native land.

Those inclined planes – how impressive in action they must have been! At Hobbacott Down, mightiest of them all, the slope was 935ft (285m) long with a vertical fall of 225ft (68m) – a gradient of approximately one in four. Here in August 1831 a boatman called John Wilkey was taking three empty boats down when somehow he allowed them to become unshackled and they crashed unchecked from the top to the bottom, causing so much damage that trade was halted for about ten days. In the south of England, only the incline on the Tavistock Canal at Morwellham was greater with a fall of 237ft (72m); there, goods were transhipped into trucks at the top of the slope, not carried down in their own wheeled vessels. The Canal and Woodland Trail in the Morwellham Open Air Museum, created on the site of the old copper port on the River Tamar, now leads to the remains of this incline and to the south portal of Morwelldown Tunnel on the Tavistock Canal.

Most of the inclined planes were operated by water-wheel, supplemented in later years by a steam-engine. At Hobbacott Down, however, the bucket-in-well system was employed, devised by the American inventor Robert Fulton, and having a fine simplicity about it. From the top of the slope two wells were dug, 225ft (69m) deep, and in each a vast bucket, 10ft (3m) in diameter and 5ft 6in (1.7m) deep, was suspended by a chain which passed over a drum which operated an endless chain that ran down the incline. When a boat arrived at the foot of the incline it would be attached to this chain, and water would then be admitted into one of the buckets, up to a maximum of 15 tons. As the bucket descended in the well, the boat was drawn up the incline; and when the bucket reached the bottom, a valve opened to release the water which flowed along an adit to the lower level of canal. Boats descending the incline would draw up a bucket from the bottom of its well. Ascending boats were usually loaded, while those descending were more likely to be empty, as sea-sand from the Bude beach was the principal cargo, distributed around the farms to lighten and fertilise the soil.

It was unfortunate that the chains for the buckets and the incline frequently gave way under the strain, and breakdowns at Hobbacott frequently brought traffic to a standstill. When Charles Brown was appointed 'Conduiter of the Plane' in 1837 he was promised a bonus of £4 a year 'provided no accident happened to the machinery by his Neglect'. He also had an assistant, on a £2 a year bonus, and the canal company insisted that neither was ever removed from Hobbacott to work elsewhere, 'there always being sufficient Work in oiling and examining the machinery when the Plane is not at Work'. The company discouraged conversation with the boatmen and decreed that 'all access by strangers to the Plane on Sundays should be prevented', which implies that Hobbacott, like Foxton locks today, was something of a weekend tourist attraction.

Only one other inclined plane in the South West was designed for the bucket-in-well system – Wellisford on the Grand Western Canal, with a vertical rise of 81ft (24.6m). Here James Green installed buckets holding 10 tons to lift 8-ton boats, but the system failed to operate. Eventually enough money was raised to replace the buckets by a steam engine, and the canal was opened.

There is something appealing in the thought of all those little boxes of goods – sea-sand, potatoes, limestone, coal, vegetables, grain – floating along narrow, shallow water channels, drawn by a horse or donkey and every now and then arriving at an incline and then turning into miniature railway trucks until the gradient was overcome. Did the boatman on the Bude, Grand Western or Torrington canals ever actually get into a boat? Or did he spend all the journey ambling along behind his horse, occasionally kicking the tub-boats out from the bank into the middle of the channel?

The Devon and Cornwall canals were quite independent, so a boatman who worked on one was extremely unlikely ever to work on another. Working on the Bude Canal seems to have been especially arduous. Samuel Parnell was a Bude Canal boatman living at North Tamerton, and according to his daughter's memories was sometimes required to work two turns a day. This meant taking a string of boats from Tamerton to the terminus at Druxton and then returning from Druxton all the way to Bude, nearly 22 miles (35km) distant. Then he would return with loaded boats to Tamerton. This took two days in all and Parnell would spend the night with his horse wherever they happened to be. Nor was he simply a boatman; loading and unloading the goods was part of his job, besides which he was also responsible for maintaining a length of canal. His wages were 14s a week, on which he kept his wife and twelve children, supplemented by some local gardening. His wife also worked at Tamerton wharf, loading farmers' waggons with coal and fertiliser.

Between Druxton and Bude there were five inclined planes – there was a sixth on the northern branch at Venn – and this brought the canal to Helebridge wharf at the foot of Marhamcurch incline, where goods were transferred from barges into the tub-boats. Parnell would therefore have had plenty of opportunity to get to know the incline keepers, each of whom lived in a canal company house by the top of his plane. Like the boatmen, the incline keepers were responsible for a length of canal; one of these, Noah Smale of Werrington, used to inspect his from the back of a pony, sitting sideways so as to get the best view.

Of the other south-western waterways, only the Chard Canal in Somerset depended on inclined planes to overcome changes in gradient. In its 13½ miles (21km) this tub-boat canal had four inclines, three tunnels – one of them, Crimson Hill, 1,800yd

Sampford Peverel on the restored section of the Grand Western Canal

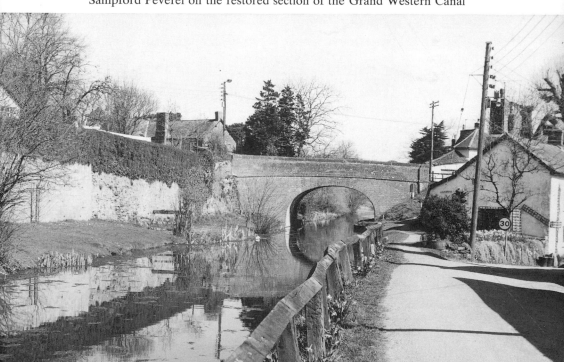

(1,645m) long – one lock and a stop-lock. The plane at Chard Common was unique: it had a single line of rails which carried a four-wheeled cradle, and a tub-boat would be loaded into this cradle. The other Chard Canal inclines were shorter and all double-track, and the boats were carried floating in caissons; at Chard Common, however, they were carried dry. One of the last of the main-line canals to be built in England, the Chard was actually under construction at the same time as the Bristol & Exeter Railway, which passed beneath it at Creech. It was this railway company that later bought the Chard Canal for £5,945 – the canal had cost about £140,000 to build – and closed it in 1867 after a working life of twenty-five years.

There is enough left of the Chard Canal to merit a good day's exploration. Near Creech St Michael is the junction of the Chard with the Bridgwater & Taunton, the Chard being carried on a substantial embankment. Off the Creech – Ruishton road is the aqueduct over the River Tone, a fine, three-arched structure but now sadly crumbling away. Opposite the Canal Inn at Wrantage a track leads to the bed of the canal and the site of the Wrantage inclined plane. Further along the north portal of Crimson Hill tunnel will be found; shackles in the tunnel roof enabled the boatmen to haul their boats along, as legging was not possible in tub-boats.

The canal passed to the west of Ilminster; Wharf Lane provides a clue to its whereabouts and there is a restored watered section on the west side of the recreation ground. Above this is Ilminster inclined plane, 82½ft (25m) long, although the tunnel at the top has been demolished.

Chard Basin was on the north side of the town, its site now absorbed into the premises of B. G. Wyatt on Furnham Road. The large reservoir created to supply the canal lies alongside the Chard–Chaffcombe road, and on its completion was described by the local paper as 'a broad expanse of surface of the deepest blue, resting sometimes in peaceful repose, and at others . . . ruffled and tossed to and fro, covered with crested waves of no mean size, its shores washed by breakers and besprinkled with angry foam.' Close by on the opposite side of the road is the site of the Chard Common incline – if you can find it. Nor is there any trace whatsoever of 'Whitelaw's patent water-mill' which provided the motive power, or indeed of any of the machinery or installations, all disposed of more than a hundred years ago.

If you want some idea of what an inclined plane would have looked like in the days of its operation in the nineteenth century, the place to go is the Ironbridge Open Air Museum by the River Severn in Shropshire. Shropshire had a small network of tub-boat canals with several inclined planes, though much has now disappeared beneath the new town of Telford. However, the Hay incline that linked the Shropshire Canal with the Coalport Canal and the wharves on the Severn has been partly restored – with a rise of 207ft (63m) it is comparable with Hobbacott Down.

James Green should be remembered as a canal engineer not only because of the inclined plane. He seems to have become a canal engineer by chance, simply because he was there – between 1808 and 1841 he was County Surveyor for Devon, an appointment which permitted him to take outside commissions. For a time he was an assistant to John Rennie, and it was probably that experience which gave him his interest in canals. Unlike other canal engineers of the time who used conventional pound locks – singly or in flights or staircases – to overcome changes in gradient, Green preferred more radical solutions, though these were better suited to tub-boats than to the 70ft (21m) narrow boats of the Midland canals. He was an adaptor rather than an inventor; inclined planes were in fact in use many decades before he introduced them to the South West, and there had been several attempts to construct boat lifts before he set to work on the Grand Western Canal.

Green was probably aware of two boat lift trials in his neighbouring county of Somerset, one of them at Combe Hay on the Somersetshire Coal Canal. Here Robert Weldon constructed his 'hydrostatick lift', which incorporated an oversized lock chamber with a lift of 46ft (14m) in which a caisson able to contain a full-size canal boat ascended and descended. After various mishaps, a successful trial took place in June 1798, and trials were repeated in the following months, some more successful than others. In what may have been the final run, more than sixty people enjoyed the experience of being lifted up and down through the water – the lock was kept full and air was pumped into the caisson in which the boat was completely enclosed. However, the masonry of the lift was beginning to give trouble and the canal committee decided to abandon the trials and try an inclined plane instead.

The second Somerset boat lift was tested in September 1800 on the Dorset & Somerset Canal at Barrow Hill. This was the creation of James Fussell, a local ironmaster and a director of the canal company, and was described by him as a 'balance lock'. Essentially it consisted of a large lock pit divided into two sections, with a watertight open box suspended in each from a framework at the top. Tub-boats would be floated into a box at the upper or lower level and would counter-balance each other as they ascended or descended. The trials at Barrow Hill were successful, the local newspaper reporting that:

> it answered the design perfectly to the satisfaction of a great number of spectators: among them many men of science, impartial and unprejudiced, who after its repeated operations and those without the least difficulty of mischance, and inspecting minutely every part of the machine, were unanimous in declaring it to be the simplest and best of all methods yet discovered for conveying boats from the different levels and for public utility.

So impressed was the canal committee that tenders were invited for five similar locks, to be constructed on the canal about half a mile to the east. But the company's financial position was deteriorating rapidly, and although excavation of the lock pits began, by the beginning of 1803 funds were exhausted and all work stopped. It is still possible to find traces of the trial lock and the unfinished lock pits on the side of Barrow Hill to the north of the railway line

James Green's final recommendation to the committee of the Grand Western Canal was for a canal with only one inclined plane but seven boat lifts, capable of handling boats of 8 tons; this proposal was accepted and construction commenced. Early in 1836, however, a notice appeared in the local press to the effect that 'Mr James Green has ceased to be Engineer to the Company', signed by the company's principal clerk. The Grand Western proved to be Green's last canal undertaking; later, his plans for the Chard Canal were rejected and no further canal building took place in the South West.

It seems to have been the failure of the inclined plane rather than anything fundamentally wrong with the lifts that was the reason for the canal company's dissatisfaction with their engineer. In general principle the lifts themselves resembled Fussell's balance lock but with more sophisticated engineering detail; they operated effectively, and what mishaps occurred were minor and easily rectified.

Waytown Tunnel, Grand Western Canal

Indeed they continued working until 1867, and it was only when the section of the canal on which they were situated – the 13½ miles (29km) between Taunton and Lowdwells – was closed that they were dismantled and the machinery sold off. Traces of some of the lift chambers can still be found, the most substantial at Nynehead, in woodland a few hundred yards west of the minor road between the A38, a mile east of Wellington, and Nynehead village. On the other side of the minor road a track leads to another memorial of the Grand Western – a fine, iron-trough, single-arch aqueduct over the Tone.

The largest of the Grand Western lifts was at Greenham, with a rise of 42ft (12.7m); the lift-keeper's cottage, now a private residence, stands above the impressive declivity where boats once rose and fell. Anyone tracing the line of the Grand Western from Taunton downwards – shallow depressions across fields, crumbling ruins here and there – might feel that this is the place to weep a last, sad tear before going home. However, carry on past the cottage and across the narrow road, and you will come to the site of a lock, a lock cottage, and 11 miles (17.6km) of water and towing path which will take you into Tiverton. This length, technically a branch of the original Grand Western, was rescued in 1966 largely by the force of local opinion and is now administered by Devonshire County Council; it is a linear park, carefully maintained and open to navigation by unpowered boats. If you seek an idyllic country canal, pottering along through villages and fields, here it is; and you can enjoy it the more from a horse-drawn narrow boat working out of Tiverton Basin in the summer months.

'An idyllic country canal': though *were* the country canals of the South West ever idyllic when they were doing what they were created for – enabling goods to be transported around the countryside? It was not boxes of rosy apples or luscious pears that filled the boats, but on the Grand Western mostly coal, culm for burning lime, lime and limestone, and in the later years, roadstone from quarries at Whipcott to Tiverton, where it was fed into a dusty steam-driven stone cracker. Between Taunton and Lowdwells the clanking and banging of the lifts and the smoke and clatter of the steam engine on the Wellisford incline would certainly have disturbed the rural peace. Even the little Torrington Canal, a 6-mile (9.6km) waterway constructed by Lord Rolle and running from Torrington to the Torridge close to Bideford, had its inclined plane at Weare Gifford, and its cargoes of coal and limestone were most unromantic. And imagine the tumultuous scene at Lusty Glaze and Trenance Point, the two terminals of the St Columb Canal where inclined planes were cut into the cliffs and boarded over. On the cliff top, boats were 'hauled up on end, and the stones were thereby shot out, and rolled down the plane to the strand below, from whence boats conveyed them on board the ships'. The stones were exported for building elsewhere, and boxes of coal and shelly sand were hauled up the inclines for distribution inland.

No: 'idyllic' is not the word as far as the working history of these canals is concerned. We tend to see the countryside of the eighteenth and nineteenth centuries through a veil of sentimentality; for a truer picture we should look at the poetry of George Crabbe and the painting of George Morland – not at the work of his engravers who prettified and tidied up his clear vision. Maybe the reason why there are so few paintings of the country canals of this period is that they did not lend themselves to the sentimentalised treatment of the more popular artists of the time.

Well worthy of mention, however, is a painting in words of a scene on the St Columb Canal, a little waterway which was never completed. It was planned by a local man, John Edyvean, as a

The northern branch of the Bude Canal at Putford in the 1960s

13-mile (21km) semi-circular canal with terminations on the coast at Trenance Point and Lusty Glaze; however, the central section was never made. Two separate lengths, totalling about 6 miles (9.6km), were all that were completed, and these operated for only about four years – by 1781 or thereabouts, the canal was disused. An abortive attempt to revive interest in the St Columb was launched in 1829 by Robert Retallick, once the superintendent of the Liskeard & Looe Canal. Traces of the canal can be found in the fields behind St Columb Minor church, including a packhorse bridge.

Our painting in words comes from *The Monthly Review* and was first quoted by Charles Hadfield. In 1779 the writer went with a group of friends to view the St Columb, and were told that its creator, John Edyvean, was on the towing-path:

> We overtook the poor old man, groping his way by the side of his canal, and leading a miserable little horse in his hand. We joined him, and he conducted us to all the parts of this ingenious work with the intelligence of one who had formed the whole, inch by inch, and this alone can account for the ease and safety with which, in his blind state, he passed through every part of it.

Edyvean was, in fact, quite an important figure in the earlier history of the south-western canals; a few years before he had been a leading promoter of a proposed Tamar Canal, to run on a circuitous course between Bude and the Tamar at Calstock. This was to be a tub-boat canal with five inclined planes, which Edyvean claimed were of his own invention. A Parliamentary Act was obtained in 1774, but the financial prospects failed to attract support. This proposal, however, engendered two others; one for the Bude Canal and another for the Tamar Manure Navigation, once envisaged as connecting with the Bude but ending up as a river improvement less than 3 miles (5km) long.

Cornwall, according to the engineer John Smeaton, 'seems but ill-adapted for the making of canals across the county, being so very frequently intersected with valleys'. The same might be said of most of Devonshire and even Somerset, and in the circumstances the tub-boat and inclined plane combination seemed the most appropriate solution. The two most financially successful south-western canals, however, were the Liskeard & Looe Junction, in south Cornwall, and the Somersetshire Coal Canal, both of which were constructed

The restored entrance lock on the Somersetshire Coal Canal

with conventional locks and navigated by vessels of more conventional size: 50ft by 10ft (15 × 3m) on the Liskeard & Looe, and canal narrow boats on the Somersetshire Coal.

The Liskeard & Looe Canal had twenty-five locks in just under 6 miles (9.6km) and was a very busy little waterway between 1830 and 1860. Coal, lime and sand were the main cargoes, joined later by copper ore from mines on Caradon Hill. In 1860 the canal company opened a railway alongside; as in South Wales, river, road, railway and canal shared the same narrow valley for some decades, and – as in South Wales – the canal was the first to go. Some of the road bridges in the valley have three arches, for river, railway and canal, and the remains of some of the locks may still be found close to the line of the railway. The penalty for swimming in the canal was one month's hard labour.

While the Liskeard & Looe Junction was a local concern, the Somersetshire Coal Canal verged on the national level. According to Joseph Priestley, writing in 1831:

This canal is of great importance in the export of coal. . . That useful article is forwarded eastward to the Kennet & Avon and Wilts & Berks canals, by which it is supplied to places on their line and also to others on the borders of the River Thames; besides entirely supplying the city of Bath and a part of the neighbourhood of Bristol.

For many years the canal carried over 100,000 tons of coal annually, peaking in 1838 with a figure of 138,403 tons, producing a revenue of over £17,000. Compare this with Bude Canal tolls of about £3,500 a year maximum, or the Grand Western which exceeded £4,000 in only three years. Indeed it was only towards the very end of its commercial life that the Somersetshire Coal Canal failed to show a profit, although the amount of tonnage carried did fall drastically in the face of railway competition from the 1870s onward. The canal went into liquidation in 1893 but failed to find a purchaser at auction the following year, and was bought in at the reserve price of £3,900. It was eventually closed to navigation in November 1898 and was sold to the Great Western Railway for £2,000 six years later. Compare this with the £30,000 paid by the Bristol & Exeter Railway for the Grand Western Canal – or with the £5,945 paid by the same railway for the Chard. Even the 1 mile 7 furlong (2.6km) Stover Canal fetched £8,000 when purchased by the Moretonhampstead & South Devon Railway. Canal property had certainly diminished in value as the years progressed.

Of all the canals of south-west England the Somersetshire Coal is the most rewarding for today's seeker after ruins and remains.* Originally 17¾ miles (28km) long with two branches, it was reduced to 10½ miles (16.8km) when the Radstock branch was converted into a tramroad, only a few years after the canal opened. However, those 10½ miles, from the junction with the Kennet & Avon Canal by Dundas Aqueduct to the basins at Paulton and Timsbury, are easy enough to follow.

The most intriguing site of all is the conjectural whereabouts of Weldon's lift. While there are various descriptions of the

---

* For a guide to almost every inch of the Somersetshire Coal canal, *The Somersetshire Coal Canal Rediscovered*, by Niall Allsop, published by Millstream Books, is strongly recommended. This is divided into several walks and includes maps and old photographs, as well as extracts from the diary of the Rev John Skinner, rector of Camerton 1803–34, valuable for conveying something of the local colour.

'Hydrostatick or Caisson-lock' in operation, and its dimensions have been carefully recorded, its exact site has yet to be discovered. Recent investigations suggest that it was close to Caisson House – once the residence of the canal's engineer – and to the site of Lock No 5 of the flight of locks which eventually replaced it. A mound topped by a chestnut tree is said to mark the place, but this has not been excavated. Maybe this fearsome device is best left to the imagination after all.

Plenty more evidence of the Somersetshire Coal Canal can be found near Combe Hay and Caisson House. Behind the house was a feeder arm of the canal connected to an engine house that pumped water back up the Combe Hay flight of locks to the canal's summit level. In front of the house is the site of the inclined plane that replaced Weldon's lock, and traces of several locks that in their turn replaced the incline. In a field to the east of the house there is a surprise: several lock chambers have been cleared out and restored by the Avon Industrial Buildings Trust, whose members have surveyed and recorded all the remains of the canal along both its lines.

The Building Trust has also renovated the aqueduct at Midford over the Cam Brook where the two arms of the canal joined. Here were wharves and sidings and buildings to cope with the coal trade, especially from the Radstock area served by the tramroad. Coal was transferred here from the tramroad trucks to boats for forwarding on to the Kennet & Avon Canal and thence west to Bath and Bristol or east towards the Wilts & Berks Canal and the Thames. There is a good turnover bridge nearby, now rescued from encroaching ivy, and the kennels on the road above were once the Boatmen's Arms.

Further along the canal on the far side of Combe Hay is the Dunkerton Aqueduct that took the canal over the Severcombe Valley, almost submerged beneath vegetation (unless the Trust has cleared it). From here westward is the old Somerset coalfield, the reason for the canal's existence, although the largest colliery, Dunkerton, was not opened until after the canal had closed. Coal was brought down the hillsides by tramroad to Radford Colliery Wharf and the basins at Paulton and Timsbury. Now, there is little which recalls the trade, the bustle, the activity and noise – some crumbling stonework, fragments of walls and buildings, the shallow depression of the canal bed. The Jolly Collier inn is a sort of memorial; as the Camerton Inn it was originally built for the canal trade, though it has now turned itself round to face the road.

If the Coal Canal is the richest in relics, another Somerset waterway, the Glastonbury Canal, is one of the poorest. Apart from its terminal basin at Highbridge, there is virtually nothing left at all of this canal which was opened with much ceremony on 15 August 1833, when two vessels 'filled with the most respectable persons in the neighbourhood' and accompanied by a band and the ringing of church bells, made their way from Glastonbury to Highbridge and back. However, the prospects of trade with the Midlands, the North and Bristol Channel ports came to naught, and twenty years later the canal was sold to the Somerset Central Railway which laid its track along much of the towing-path.

When the canal was opened, Glastonbury was a busy market town with a ruined abbey and several fine old inns; it is now a 'heritage centre' too, and a focus for New Age seekers for Truth. You can drink from the Chalice Well, climb the Tor, buy any amount of crystal – but there is no sign of the Glastonbury Canal. Derelict canals in flat country soon vanish – the Caistor Canal in Lincolnshire is another example. Traffic on the Glastonbury Canal disappeared some years before it was closed; at a meeting in 1848 when purchase of the canal was being discussed, one railway shareholder wondered whether anyone had ever seen any boats or traffic on the waterway. 'I have been there time after time,' he said, 'and have never seen a boat in my life.'

Elsewhere in the south-western counties, however, a number of 'secret places' still survive, relics of old navigations that have somehow escaped destruction and which speak as eloquently of the past as any historical monument. One such is Murtry Aqueduct on the line of the Dorset & Somerset Canal near Frome, a three-arch stone structure over a stream built in 1798 but never used by boats. It is about a mile (1.6km) east of the unfinished lock pits on Barrow Hill. Through the copse in which the aqueduct hides there is a length of canal bed, lying between the aqueduct and the Frome–Buckland Dinham road. Like the taller twin-arch Coleford Aqueduct on the canal line to the west (described as 'noble and stupendous' at the time), Murtry Aqueduct is a monument to hopes unrealised; not hopes for a little branch canal between Nettlebridge and Frome, but for a mighty waterway some 48 miles (77km) long from the Kennet & Avon to a point in south Dorset between Sturminster Newton and Blandford Forum, with a branch from Nettlebridge intended to carry coal from the Mendip coalfield. The estimated cost for this was £115,400, with a further £30,584 for the branch. But in the event, the few miles of the branch that *were* made cost the canal company over £66,000, and not a penny of revenue was ever raised.

Another aqueduct hides away in North Devon, disguised now as a stately bridge over the River Torridge, part of the drive to a manor house. This is Rolle or Beam Aqueduct, and the Torrington Canal, which it was designed to carry across the course of the river, featured in one of the most moving country books ever written: Henry Williamson's *Tarka the Otter*. It was the private venture of an influential local landowner, Lord Rolle, who presumably had some say in the classical design of the impressive five-arch aqueduct; James Green was engineer. It was by far the most substantial and ambitious of the half-dozen other private canal undertakings in the South West; the 6 mile (9.6km) Torrington Canal included an inclined plane and a tide-lock as well as the aqueduct, while none of the others exceeded 2 miles (3km) and only the Stover Canal had any engineering work of note.

Rolle or Beam Aqueduct on the Torrington Canal. This is Canal Bridge in Henry Williamson's *Tarka the Otter*

Commercially, however, the Stover Canal was far more successful. It was created by James Templer of Teigngrace for the clay trade; clay dug locally was carried in 54ft (16.4m) barges through five locks to the River Teign and exported from Teignmouth, much of it finding its way to Staffordshire. Imports included coal, limestone and sea-sand. When the Haytor Granite Tramroad was opened in 1820, granite was, for a time, added to the canal's cargoes, the stone being shipped to London for various public buildings including the British Museum and the National Gallery. In 1867 the top section of the Stover Canal was closed by the Moretonhampstead & South Devon Railway, which had bought the canal for £8,000, but the remainder continued in use until 1939. A 'Heritage Trail' at Teigngrace leads to Teigngrace Lock and all sections of the Stover Canal.

The heritage trails, theme parks and so on which have opened in more recent times are a clear indication of renewed interest in the countryside and the past, and also reflect how much more leisure time and mobility the town-dweller enjoys. So do open-air museums, and the one at Morwellham on the River Tamar has introduced a great many people to the Tavistock Canal. Nowadays it is easy to inspect both the inclined plane and the south portal of the 2,540yd (2,298m) Morwellham Tunnel; on 24 June 1817 several hundred people 'of all ranks' were conveyed through the tunnel in nine iron boats 'under a salute of banners bearing appropriate inscriptions together with a party of miners and others dressed uniformly with ribbons in their hats having on them the words 'Success to the Tavistock Canal'.' As this was one of the smallest bore tunnels ever built, 7ft (2.1m) wide and 12ft (3.6m) from arch to invert, the banners must have been modest.

The Tavistock Canal took thirteen years to build and cost a great deal more than its estimate. It was created to serve the copper mines, but was opened too late to realise the trade that its originators had expected, although it dragged on an unprofitable existence for over sixty years. It can be followed from the car park at the end of Canal Street in Tavistock, once Tavistock Wharf. The canal is in water as it supplies a generating station above Morwellham; the towing-path takes you across the Lumburn Aqueduct, past the junction of the Mill Hill branch – which you can follow if you wish – and on to the north portal of the tunnel.

What else does the South West have to offer? The sea-lock at Bude, and its wharves – look out for the narrow boat here, now a

Maunsel Lock on the Bridgwater & Taunton Canal (British Waterways Board)

Christian bookshop, which arrived not via the tub-boat length nor by sea, but on a low-loader from Birmingham; also the three-storey warehouse at Westport by the terminus of the Westport Canal; the Exeter Ship Canal – the first purely navigational canal to be built in Britain – with the massive Double Locks, its ferries, the Turf Hotel and the maritime museum, where the least spectacular exhibit is a Bude tub-boat with wheels; the single lock of the Tamar Manure Navigation near Gunnislake, now a scheduled ancient monument; and of course the whole of the Bridgwater & Taunton Canal, one of the very few south-western undertakings that was completed more or less according to plan.

As the only complete and surviving inland navigation in the South West the Bridgwater & Taunton is somewhat disappointing. It seems very much a poor relation of the canal system: isolated, little used except by light craft and trailed boats as low bridges severely limit headroom, and lacking in buildings or engineering structures to attract the attention of the visitor or passer-by. Nevertheless, in its day it was an undertaking of some power and importance; it completely overwhelmed the Tone Navigation and took it over, and it was responsible for the development and prosperity of Bridgwater docks. It connected with the Grand Western and Chard canals and the rivers Tone and Parret. Its best years were 1842 and 1843 when it carried over 100,000 tons with revenue running into five figures. Bought by the Bristol & Exeter Railway, whose line between Bridgwater and Taunton was constructed nearly parallel to the canal, it fetched £64,000, no mean sum when compared to the price of other canal undertakings.

For about ten years the Bristol & Exeter kept it going, even investing in new locks and bridges; and only when that company amalgamated with the Great Western Railway did decline really set in. Commercial traffic ended early in the twentieth century and since then the canal has been neglected. Wartime brought a flicker of interest when the canal was envisaged as a defence line against invasion; a legacy of pill-boxes attests to this. Then in 1966 the Somerset Inland Waterways Society was formed dedicated to work for the restoration of the canal, and some interest was re-awakened. A measure of life has returned to the waterway and a restoration programme is in progress.

The Bridgwater & Taunton Canal was built with several swing bridges, most of which were fixed during the war when the military destroyed their mechanism. Some of these have been rebuilt, but the headroom of 4ft (1.2m) is not encouraging for larger boats. Much of the countryside on the route has been urbanised, although some pleasant – if undistinguished – stretches still remain. Maunsel lower lock with its unique paddle mechanism is the canal's most distinctive feature, carefully restored and in a beautiful setting. The canal dock at Bridgwater, and Bridgwater Docks altogether, deserve both exploration and revival; maybe the restoration of Gloucester Docks will provide some inspiration.

Existence was always a struggle for the canals of south-west England; prosperity was seldom achieved and then only briefly. For almost all of them – one-half of the Grand Western and the Bridgwater & Taunton are the exceptions – the 'leisure revolution' came far too late; by the time the amenity value of waterways was recognised, these country canals had long since been mouldering away, their lifts demolished and their inclines overgrown. They had reverted to being part of the countryside – part of the fields to which they had once brought fertiliser, and whose produce they had carried away. Now they are as distant as the drovers' tracks and green lanes of the far-off past.

Ramsey in the early nineteenth century (see page 91)

8 Over the Hills and Far Away

For most of us, 'a canal' means the narrow canals of the Midlands, centred on Birmingham or Hawkesbury Junction – narrow canals with narrow locks for narrow, brightly painted boats. Occasionally a full-length boat passes by, but it is rarely a working boat, and very rarely a working pair which would be a sight to be photographed and treasured. Every few miles there is a boatyard with craft for hire, a shop stocking painted ware, guides and confectionery, a fuel supply and pump-out station. There are cheerful canalside pubs, quaint hump-backed bridges and canalside cottages with carefully tended gardens – all as pretty as a picture.

But the Midland narrow canals – Shropshire Union, Trent & Mersey, Staffordshire & Worcestershire, Worcester & Birmingham, Coventry, Oxford and the rest – are only a part of our inland waterways. Elsewhere many of our country canals did not touch any of the larger towns except possibly at a terminus, and in their time served predominantly rural areas – the Yorkshire Wolds, the Welsh Border counties, Cotswold villages, mid-Wales. Many of them are indeed over the hills and far away. Some are long abandoned and derelict, others are being restored, but several are very much alive – like the Llangollen Canal or the Brecknock & Abergavenny – crowded with boats, in the summer as busy as ever they were in the days of their commercial activity.

Several of these canals were linked with river navigations and were designed to take the same craft: the Pocklington Canal took Humber keels, the Louth Canal in Lincolnshire sloops and sailing barges, and the western section of the Thames & Severn, Severn trows. Canals linked to the main network could take full-length narrow boats, although these would have had to be especially slender to navigate the Leominster Canal as the locks were less than 7ft (2m) wide. Isolated canals had their own unique dimensions – for example the Brecknock & Abergavenny and the Monmouthshire which were 64ft 9in (20m) by 9ft 2in (2.9m); boats on these two canals had neither bow nor stern, and the rudder was simply

A scenic canalscape. Shipton-on-Cherwell church on the Oxford Canal (British Waterways Board)

transferred from one end to the other. Many of the country canals could be navigated throughout in less than two days, so there was no need for family boats or craft with cabins. Painted boats were strangers to nearly all the rural waterways, except for the Llangollen and Montgomeryshire canals and the Wilts & Berks. A few narrow boats off the Nottingham Canal crossed the Trent to the Grantham Canal carrying mostly coal, bringing back agricultural produce from the farms of Lincolnshire.

The Grantham Canal looks more like a river as it meanders through the gently rolling countryside of the Vale of Belvoir. It potters along from village to village, linking them rather than avoiding them, and fits comfortably into the landscape. All it needs are boats; but for boats to return locks need to be repaired and bridges with adequate headroom rebuilt. At the Grantham end a new basin would be required, perhaps on the west side of the A1, and at the western end a new connection with the Trent. It sounds a lot, but compared with what has been achieved on, for example, the Kennet & Avon it is well within possibility. Restoration of the Grantham Canal – what a proposal for sponsorship by one of the larger Nottingham breweries that would be!

Restoration is in hand on many rural waterways, including the Montgomeryshire, the Pocklington, the Thames & Severn, and the Wey & Arun Junction. At first inspired by local interest and the enthusiasm of voluntary societies – the Inland Waterways Association, the Waterway Recovery Group – proposals eventually attract the support of the British Waterways Board and the local government authorities. Some organisation may provide a work force and the EEC may give a grant. Mile by mile, lock by lock, work proceeds, until ahead appears a major obstacle: Sapperton Tunnel, the locks in Sidney Wood, the water supply for a summit level. Politics may intervene; the Secretary of State for Wales omits the Montgomeryshire Canal from a list of projects for EEC grant aid, or local politicians may withhold their support, as happened for years with the Kennet & Avon scheme. Nevertheless, things do get done: the Southern Stratford Canal has long since returned to operation; the Kennet & Avon is fully operational, and the Basingstoke Canal nearly so; and boats have returned to the Herefordshire & Gloucestershire Canal, even if they do have only about half a mile (0.8km) of clear water.

Perhaps the restoration of the Thames & Severn is the project with the greatest possibilities. Planned as a vital link in an east–west waterways route linking London with the ports on the River Severn, the canal was cut from a junction with the Stroudwater Canal at Stroud to the Thames above Lechlade, 29 miles (46km) away. Apart from a branch to Cirencester it passes through no towns and few villages – it traverses the fertile countryside of the Golden Valley until it climbs to its summit level, then plunges into Sapperton Tunnel to emerge in the Cotswold uplands, a bleaker landscape. Thence it descends to the Thames, meeting it near the river's head of navigation.

In its trading life the line of the Thames & Severn was maintained by twelve watchmen, who looked after the locks on their patch and 2 or 3 miles (3–4km) of canal, towing-path and hedge. The watchmen would also keep an eye on the boatmen, reporting to the company any breach of regulations. They worked long hours: sixteen hours a day in summer, twelve in winter; for this they were paid 9s (45p) a week, and provided with a company cottage.

Of these cottages, five – all surviving – were 'round houses':

circular dwellings on three storeys resembling miniature tubby light-houses. Imagine what the scene might have been when, for example, an applicant for the job of watchman at Coates visited the round house with his fiancée. The situation could very well have developed as follows: first there is the walk along the towing-path from Tarlton Bridge, with the gloomy east portal of Sapperton Tunnel in the cutting on the far side. Then the isolated round house with its narrow lancet windows and its front door up a flight of steps on the first floor. There is a side door at ground level; this is unlocked, and the happy couple, perhaps a little apprehensive, open it. Inside it is dark and the air is steamy. It smells of . . . 'Donkey!' she cries. She is right – this is the donkey's stable, circular and nearly 17ft (5m) in diameter. Against the wall, a staircase leads upwards to the living-room, usually entered from the front door in the opposite wall. This room is also dark, with just two narrow windows looking on to the canal in each direction. On one wall is a kitchen range.

The couple exchange glances. The staircase continues upwards and here is the bedroom, waiting to receive the matrimonial bed – if there is any way in which it can be introduced. It won't go through the narrow windows and it would seem impossible to manoeuvre it up the stairs which are too narrow and awkward for any furniture except small chairs and stools.

That is about all. Outside is a lean-to shed and a privy a few yards away; furthermore, apart from a well that looks both deep and dry, there is no visible water supply. As they stand there, a horse-drawn boat appears, a young lad leading the horse, a burly, middle-aged man wearing a plush waistcoat, moleskin trousers and a cap at the tiller. The boat comes to a halt in the stop-lock outside the house, although the gate is open. The boatman steps off, rummages in the patch of rough garden and returns with a cabbage and a few carrots. 'That's about the last of it!' he shouts to his unseen wife in the cabin. 'Won't be any more here till they get a new fellow in. Take 'er on, Jim!' The lad slaps the horse back into action, and off they go.

'We'll need a strong fence,' says the new watchman, 'and a proper gate.' His fiancée is less accommodating: 'Who says we're coming here at all? This mucky, dark place with no furniture and a smelly stable is *no* place to bring up children!'

No-one else had ever complained, the canal company agent at Siddington tells them next day. 'Well, I am!' says the watchman's fiancée. 'Look at those nice houses at Sapperton and Furzen Leaze. Why can't you build us one like that?'

The approach to Sapperton Tunnel in the early days of the Thames & Severn Canal

The west portal of Sapperton Tunnel on the Thames & Severn Canal, 1986

And eventually she won her point: the Siddington agent consulted the clerk of the works, who consulted the clerk to the company, and within three weeks workmen appeared at the Coates round house to make the ground floor into a living-room and build a new stable outside, to add a separate kitchen and another wider staircase, and to partition the bedroom. The couple moved in.

The company, however, was running into major difficulties. Leakage from the summit level had always been a problem, but with falling revenues, mainly because traffic was deserting the canal for the railway, maintenance standards were deteriorating, and at times there was barely enough money to pay wages. In 1893, a few years after the couple moved in, the canal east of Chalford was closed. Efforts were made to save the canal and the closed length was reopened six years later, but the leakage and the financial difficulties remained, and traffic over the summit level soon ceased completely. In 1933 the canal was legally abandoned.

Sapperton Tunnel, 3,187yd (2,914m) long and the third longest canal tunnel in Britain, is the main obstacle to restoration of the Thames & Severn. It has left its mark on the locality demographically as well as geographically. Between two and three hundred men were employed in digging the tunnel, including miners from Derbyshire, Cornwall, Somerset and Wales. Many of them lodged in the New Inn – now the Tunnel House – where the first and second floors were originally dormitories; others stayed in the Bricklayer's Arms – now the Daneway Inn – near the western portal. Some married men moved into the villages of Coates and Sapperton; other men married local girls. Births, marriages and deaths all increased during the five years it took to complete the tunnel, several of the deaths due to accidents in its construction. Yet no disaster at Sapperton quite equals the tragedy at Southnet Tunnel on the Leominster Canal where two men and a boy were entombed in a collapse; their bodies were never recovered, and Southnet was never used by boats.

Sapperton, Southnet, Oxenhall and Putnal Fields on the Herefordshire & Gloucestershire: a tunnel gives an added dimension, sometimes even a touch of glamour, to a canal. Working boatmen found them tiresome as passage was slow and usually wet, and could be dangerous, especially in the early days of coal-fired engines when choking fumes quickly accumulated. However, to pleasure boaters a voyage through a long tunnel is an exciting adventure, especially for the first time.

'Into the grim darkness you glide and, within half an hour, are lost in a sightless cavern where the drip drip of the clammy water sounds incessantly in your ears,' wrote Temple Thurston of his voyage through Sapperton in the *Flower of Gloucester*. Often there is a sense of relief at surviving the ordeal: 'When we emerged at the other end we really appeared to be floating on air,' said Harold Schofield after paddling through Sapperton in 1868.

One day Sapperton may be reopened; the Coates portal has been restored and restoration of the whole length no longer appears an impossibility. Tunnels on disused canals usually just decay; the channel silts up, the invert rises, parts of the roof fall in. The roots of trees and shrubs infiltrate the stonework of the portals and the ventilation shafts collapse. For those who love ruins the portals of Oxenhall Tunnel, the deep cutting leading to Ashperton Tunnel – both on the Hereford & Gloucester Canal – would prove especially rewarding; so would the east portal of Southnet, though the other has long since disappeared.

'If you are fond of water and of England, to wander along stretches of the canal or to work it by boat will satisfy the most ardent countryman,' wrote General Sir Hugh Stockwell in 1977. He was writing about the Kennet & Avon, of whose Restoration Trust he was the leader for many years. He goes on to describe its route:

> The canal crosses the low-lying country between Reading and Newbury, soft marshland covered with rushes and full of wildlife; the country climbing away to the south over the lush moorlands of Berkshire, past the mellow grasslands of Hungerford to climb away to the west, reaching the summit level by Crofton; then along the southern slopes of Savernake Forest, along under the chalk downs of Wiltshire – areas of the earliest English settlers, and later the Saxon villages, unspoilt, uncluttered and peaceful; along the canal's longest lock-free stretch from Wootton Rivers to Devizes, some 15½ miles [25km]. It then runs down into the Wiltshire Vale, rich farmlands, to Bradford-on-Avon and Bath, full of lovely soft stone buildings and steeped in history. The Kennet & Avon finally completes its journey to Bristol along 11 miles [17.6km] of the River Avon.

General Stockwell died before 'his' canal was reopened throughout: appropriately he was succeeded in the leadership of the Trust by an admiral. The General – who, in his own words, 'lost one damned

The Barge, Honey Street Wharf, on the Kennet & Avon Canal

canal at Suez and didn't intend to lose another' – is commemorated in the name of the restored top lock of the Caen Hill flight at Devizes. The Kennet & Avon Canal was formally reopened by the Queen in the summer of 1990.

Crofton and Claverton pumping stations, Dundas and Avoncliffe aqueducts, the Caen Hill flight, the Devizes canal museum: all these attract thousands of visitors annually. However, perhaps more characteristic of the Kennet & Avon as a country canal is the small canalside village of Honey Street, between Devizes and Pewsey. It has neither church nor chapel, and owes its existence entirely to the canal; in its time it played an important part in the canal's history. Where the neighbouring villages lie away from the canal, Honey Street nestles close beside it, and its inn, the Barge, is the only waterside pub between the Black Horse at Devizes and the French Horn at Pewsey, nearly 12 miles (19km) apart.

Honey Street wharf dates from 1811, shortly after the canal was opened. The first Barge Inn was built about this time; a stone on the wall records that it was destroyed by fire on 12 December 1858, and it was rebuilt in six months by Samuel Robbins, with Ben Biggs as the architect. The Barge brewed its own beer, baked its own bread and ran its own butchery for the canal community. Samuel Robbins was a member of the Robbins family who established a boat-building and trading business at the wharf. Robbins & Co brought in timber from Avonmouth and various places along the canal in their own boats, storing it in their timber yard and using it to build barges, not only for the Kennet & Avon but also for the Basingstoke Canal, the Wey Navigation and the River Avon at Bristol. Some were known as Kennet barges and the boats of one class, built for trading on the Avon, were named after precious stones – *Emerald, Pearl, Ruby* and so on. One or two of these may still survive.

In later years Robbins & Co became the firm of Robbins, Lane and Pinnegar Ltd; they continued trading into the 1930s and were the last regular carriers to use the canal. It was lack of maintenance of the waterway and the dilapidation of the works under the canal's then owners, the Great Western Railway, that compelled them to close down. In 1950, however, conditions seemed good enough for an ex-naval commander to set up a new business at Honey Street: a boat-hire firm, grandly named 'The Sierra Line'. It consisted of four converted pontoons, one with an outboard motor, and one fitted with sternwheel propulsion which coped quite successfully with the accumulation of weed in the canal. For a year the Commander struggled on; then he sold three of his boats, handing over the sternwheel craft, the *Wayfarer*, to John Gould, who was himself fighting a long battle to keep the canal alive.

Many of the early wharf buildings and cottages survive at Honey Street; the Canal Trust's vessel *Charlotte Dundas* still uses the wharf as its base, although the bustle and activity of the boatyard have disappeared. The history of this site is even older, for when the Kennet & Avon was cut it bisected the line of one of the most ancient roadways in Europe, the Ridgeway, just to the west of Honey Street. The canal company provided a ferry and ferryman to complete the right-of-way; later, the ferryman was withdrawn though the boat remained for travellers to use. Still later a floating bridge was provided, which had to be punted across. Now, however, you have to cross the canal by the road bridge on the east side of the wharf.

In the southern half of England the Kennet & Avon and the Thames & Severn were the two great cross-country routes. Between them, running north-eastwards across Wiltshire and Berkshire from a junction with the Kennet & Avon near Melksham to the River Thames at Abingdon, was a sort of poor relation: the narrow-locked

Wilts & Berks Canal. This wandered for 51 miles (82km) across country, with branches here and there to places that in the canal's lifetime were no more than large villages – Calne, Chippenham and Wantage. The longest branch, dignified by the name of the North Wilts Canal, linked the Wilts & Berks with the Thames & Severn Canal at Latton, near Cricklade. One of the villages close to the main line of the canal was Swindon, with a population in 1801 of just under 1,200. A milestone, 'Semington 26 miles' in the Parade Shopping Centre, is one of the few reminders left.

Although the Wilts & Berks operated for over a hundred years there is now little evidence of its existence on the ground; like several of the canals of the South West much of its line has been completely ploughed back into the fields through which it once carried boats. Never a profitable undertaking, it did, nevertheless, in the words of its historian L. J. Dalby 'contribute materially to the prosperity and development of northern Wiltshire and Berkshire, providing the upper Avon valley and the Vale of the White Horse with a cheap supply of coal'. Much of this coal came from the Somersetshire coalfield and it was the failure of this, together with the competition from the Great Western Railway, that caused its demise.

Throughout the nineteenth century London was linked to the Severn estuary by three waterway routes which also served the towns and villages of Berkshire, Wiltshire and parts of Somerset and Gloucestershire. There was also a link with the south coast via the Thames, the Wey, the Wey & Arun Junction Canal, the Arun Navigation and the Portsmouth & Arundel Canal: a voyage of 116 miles (186.5km) with fifty-two locks. Between 1824 and 1826 there was a monthly bullion run between Portsea and the Bank of England, an armed guard accompanying the barges on their four-day voyage. In the following years this particular traffic fell away to one voyage every six months or so, and in 1838 it ended altogether. By this time there was very little other traffic on the Portsmouth & Arundel Canal, and in 1839 a local directory commented that it was 'of little advantage to the inhabitants, this being an agricultural district'. Yet, as the canal's historian, P. A. L. Vine says 'one of the reasons advanced only twenty years before for building the canal had been to improve agriculture! Indeed,' he continues, 'the Portsmouth & Arundel Canal had been a stupendous failure, costing more pounds to build than tons it was ever to carry throughout its length.'

The Wey & Arun Junction struggled on into the 1870s. It was handicapped throughout its lifetime by shortage of water at the summit, due partly to inadequate supply and partly to the large number of locks, and it was in no shape to cope with competition from the railway. Its route is wholly rural and nearly all of it can be followed today, thanks to the Wey & Arun Canal Trust which is carrying out restoration work at points along its length. Paul Vine quotes an article from the *West Sussex Gazette*, June 1866, describing a voyage along it:

> Without exception the scenery throughout is the most charming and delightful ever witnessed. . . . The glory of the scenery is its rural simplicity and its richness. At every bend of the canal you see a picture ready for the artist. The landscape is not distorted by cockneyfied villas, or 'villers' as they are called in derision – indeed you see but a few old farm houses on the journey, and they are very old and picturesque indeed. What the population may be surrounding the 18 miles of canal we know not, but we should hardly suppose it to exceed 2,000 or 3,000, taking some distance from its banks. It lies in a purely agricultural district, and a large quantity of woodland flanks the canal on both sides, and here grows some fine oak. As far as local trading is concerned the trading would consist of timber and bark. In a passage through this delightful district one almost fancies himself in an unexplored country, away from the haunts of man.

Today you may find some 'villers' here and there but much of the route still accords with this description, especially in and around Sidney Wood.

Other southern country canals fared no better than the Wey & Arun. In Hampshire the Andover Canal was abandoned in 1858, while the uncompleted Salisbury & Southampton lasted for only three years. In Kent the Thames & Medway Canal converted itself into a railway, and track was laid through the 3,946yd (3,608m) tunnel at Strood; passengers are still carried through it today. The only canal to survive as a water channel – though not as an active navigation – is the Royal Military Canal; it was also the only one built by the government as a defence work, a sort of watery Maginot Line. With a French invasion threatened on the resumption of war in 1803, the canal, with a road alongside, was constructed on the high ground behind Romney Marsh, where it was considered an invasion force might land; the Duke of York, commander-in-chief, said of the canal that it 'may be fairly considered as an almost insurmountable

Barrier against an Enemy's penetrating into the Country'. However, by the time it was finished Napoleon's eyes had turned towards the east, so the government then had to make the canal earn its keep.

The first step was to appoint most of the great officers of state as commissioners to be responsible for the maintenance of the canal and to promote its trade. Stone, timber, bricks and sand for local use were the main cargoes, carried in barges owned by the commissioners and towed at times by horses of the royal waggon train. Tolls, however, never amounted to more than a few hundred pounds a year, so the Royal Military Canal devolved into a country canal serving a moderately useful but not especially profitable purpose in moving heavy cargoes across the Kent countryside. Evidence of its original military purpose can still be seen in the zigzags in its line, made so that the defending guns could fire along each stretch without hitting the defenders. Like the Martello towers built at the same time, it stands as a reminder of past wars.

Many country canals were the result of the so-called 'Canal Mania' of the 1790s, when the success of the older canals seemed assured and it appeared to the promoters that profits were there for the taking. Among these were the Brecknock & Abergavenny,

Alvingham lock chamber on the Louth Canal

the Grand Western, the Leominster and the ill-fated Dorset &
Somerset. Lincolnshire, however, already had two country canals;
one of these was the Stamford Canal, a 9½ mile (15km) locked
waterway designed to replace the navigation of the River Welland
between Stamford and Market Deeping. This was the creation of
a Stamford alderman, Daniel Wigmore, who made it at his own
expense in return for a lease of the tolls. It lasted until 1863 and by
importing sea coal and promoting the malt trade it played its part in
the prosperity of that most handsome town.

The other early Lincolnshire navigation was the Louth Canal, a
wide waterway between Louth and Tetney Haven on the Humber
estuary. The first survey for this canal took place in 1756, three
years before the Act for the Bridgewater Canal, although no-one
seemed in any hurry to get on with matters – one engineer's report
was delayed until the spelling had been corrected. Work began in
1765 and took five years to complete.

The Louth Canal was closed to navigation in 1924 although it is
still a water channel. It penetrates a remote corner of England; a flat
and lonely land, 'the level waste, the rounding gray' as Tennyson,
who went to school in Louth, described it in *Mariana*. Through this
level waste – not quite level, as eight locks were built on the canal –
sloops and sailing barges carried cargoes of coal, grain, timber and
groceries, up to 120 tons. Six of the locks, all of slightly different
dimensions,were made with four bays on each side with wooden ties
where they met, the intention being to strengthen them against the
pressure of the surrounding land. As long ago as 1895 a proposal was
made by Mr T.R. Matthews to the canal company that the Louth
Canal should become a recreational waterway:

> A pretty little steamboat for pleasure trips to Louth and
> back would be, I think, a decided success. A few nice boats
> for rowing parties would form a profitable investment, and
> why not let an acre or two of land and make the Tetney Lock
> Canal Gardens which would make the mouths of Sheffield
> people water?

Why not indeed? It's still a good idea.

Most of Lincolnshire's navigable waterways were either rivers or
river improvements, like the Horncastle and Sleaford navigations, or
navigable drains – there are still, for example, over 50 miles (80km)
of Witham Navigable Drains open to navigation. An attempt was

Driffield. The head of navigation

Church Bridge, Thornton, on the Pocklington Canal

made, too, to link the market town of Caistor to the River Ancholme by means of a locked canal, but the Caistor Canal ended 4 miles (6.4km) short of its target and has left little mark on the landscape. And the Alford Canal was never even begun, despite appearing on various early maps of inland navigation.

North of the Humber estuary, in the Yorkshire Wolds, are four more country canals: the Pocklington, Driffield, Market Weighton and Leven. North-east of Beverley there are miles of flat land, almost fenny in character, drained by dykes and becks, most of which feed into the River Hull. In the late eighteenth century much of this land was owned by the Bethell family, agricultural improvers, land drainers and turnpike road promoters, and it was Mrs Charlotte Bethell who created her own 3¼ mile (5km) canal between Leven and the Hull. Mrs Bethell was one of those few individuals who succeeded in promoting, building and running their own canals – among others were the Duke of Sutherland, Sir John Ramsden and the Daniel Wigmore of Stamford we have already met. She and her family were more successful than most; her canal was generally profitable and it was owned by the family until 1963 when, closed for some years to navigation, it was sold for just under £2,000. A few small boats from a caravan park on its banks still bob about on its water.

Another Yorkshire lady whose name is linked with a canal is Mrs Sheila Nix, secretary of the Pocklington Canal Amenity Society for over twenty years. In 1969 it was proposed that this fine waterway – then no longer navigable – should be used for the dumping of sewage treatment sludge, and only the intervention of local enthusiasts saved it from this fate. Today the Pocklington Canal is being actively restored; since 1987 boats have been able to reach Melbourne Basin from the River Derwent, and a few more years will see the canal reopened throughout.

The Wolds waterways are essentially rural. Neither the Pocklington nor the Market Weighton canal has a terminus in either namesake; the Driffield Navigation stops at the edge of Great Driffield, and the basin of the Leven Canal is well to the south of the little town. Canal Head is a good mile south of Pocklington; it is now a quiet picnic area with the old warehouse by the basin converted into flats, and the lengthsman's hut occupied by the Amenity Society, sometimes by Mrs Nix herself, ready to answer visitors' questions. It is nearly sixty years since the last visit of Mr J. W. Brown's keel *Ebenezer* with a cargo of roadstone, most of which had to be offloaded into lighters before Canal Head was reached. The Pocklington Beck supplied the summit level at Canal Head with water, but it used to bring in a quantity of silt as well which necessitated frequent dredging; then as trade and revenue decreased, so did the frequency. Mr Brown became tired of sitting in *Ebenezer*, which in turn sat on the mud, and frustrated at having to offload so much of his cargo half-way along the canal. Soon afterwards he sold his boat and bought a lorry.

The outstanding feature of the Pocklington Canal is its brick bridges. Late Georgian, with flowing lines and substantial wing walls, they are unique among canal overbridges – though not popular with motorists owing to the exaggerated and vision-obscuring hump. All the same, they span the canal with a majestic confidence. The Pocklington is also the only East Yorkshire canal connected to another navigation, and it will become more valuable if and when the navigation rights of the upper Derwent are recognised.

The East Yorkshire waterways served no towns of any size, and did well to survive commercially as long as they did – all of them well into the present century. The Driffield Navigation is likely to be reopened throughout fairly soon, too, so that 'once again a Yorkshire keel will be able to navigate to Drillfield', if any can be found. Hopefully it will be easier to find keels than it was to find commissioners a few years ago, when none of the ninety-five appointed to administer the navigation could be traced. In a way this was not surprising, as they had originally been appointed by a Parliamentary Act of 1767; but in the absence of any of their descendants or nominees no-one had the authority to take decisions. And when Wansford Bridge needed rebuilding, the acting clerk to the commission consented to a plan that allowed inadequate headroom for navigation. However, in 1974 new commissioners were appointed – only fifteen this time, but with the positive intention of reopening to navigation.

Canal Head is one of the features of Great Driffield, a wide, deep basin with handsome warehouses and two well-preserved hand-operated cranes. The warehouses have been tidily converted into flats and the whole area is neat and well-kept. Outside Driffield the navigation resembles a village canal, running alongside the road through Wansford, with both trout and a Trout Inn, and with moorings at Brigham, a sailing club and a scow club. With their 200sq ft (18sq m) of sail, the scows can tack quickly in the narrow waterway.

Last of the East Yorkshire navigations is the Market Weighton Canal. It resembles a Lincolnshire navigable drain, running in almost a straight line through flat and mainly featureless countryside. The southernmost 3 miles (5km) or so from Weighton Lock on the Humber to Newport, are still open to boats. With its terminal basin on a minor road 2 miles (3km) from Market Weighton and with no village or settlement on its route, the possibilities of trade were limited – Newport in fact grew up where the canal was crossed by a road and later by railway. But the canal was cheap to construct, and with only four locks seems to have been easy to maintain; for a few years it even paid a small dividend. Recently it has gathered some supporters, but as it is isolated and difficult of access it seems unlikely that the Market Weighton Canal will capture the imagination of today's pleasure boaters. Place-name enthusiasts might enjoy it, though: along or close to its line are Eight and Forty, North America, Land of Nod and Duck Nest, while a little further away are Ladies' Parlour, Snake Hall, Bunny Hill and Rascal Moor. Possibly, however, the canal is better characterised in the name of Sod House Lock.

These little East Yorkshire waterways contributed to a degree to the general prosperity of the area – that they survive, and that their future seems assured, demonstrates the affection felt for them by local inhabitants and their value for walkers, anglers, nature-lovers and canoeists. Conservationists sometimes oppose canal restoration in the mistaken belief that once a rural waterway is navigable it will become infested with powerful cruisers like a Thames lock on a summer Sunday. But as Sheila Nix says '. . . the canal can provide for a wide variety of interests, and canal boats and wildlife can co-exist happily together. We believe in conservation of wildlife and conservation of the waterway and its working structures. We know that both give pleasure to a great many people.' The evidence from the canals and river navigations that have been restored supports this, as can be seen from the healthy condition of most canal fisheries – the best indication that all is well with the natural life of the waterway.

The 'canal mania' of the 1790s was not responsible for any of the East Yorkshire country canals, but has found an echo in the 'restoration mania' of recent times. The Herefordshire & Gloucestershire Canal was a child of the first and a subject of the second; it was built in two instalments, but the money ran out when only 16 miles (24.6km) – running roughly northwards from Gloucester – had been completed. Then in 1827, nineteen years after work had ceased, Stephen Ballard was appointed clerk to the canal company. His enthusiasm drove the directors to agree to complete the line, and in 1845 the canal to Hereford was open with twenty-two locks, three tunnels – and little traffic.

The railways came late to Hereford, which meant that the canal still had some purpose; it was most useful in serving the villages and small market towns of Ledbury and Newent along its route. Coal mines near Newent were meant to bring it prosperity, but they were soon exhausted. Lock cottages and wharf buildings still dot the countryside, severe, red-brick structures, easy to recognise. In the 1860s the canal was leased to the railway, and in 1883 it was closed, the Gloucester–Newent railway being built on part of its line. The canal company continued in existence until 1945, and Archie Ballard, grandson of Stephen, remembers his father and three brothers going every year to the Bell at Gloucester for dinner and a company meeting. 'The railway hadn't been able to afford to buy the canal,' he said, 'so they paid the company £5,000 a year not to use it and every year my father and his brothers held a meeting

Lost beyond recall? The Herefordshire & Gloucestershire Canal, near Oxenhall, 1987

and divided up the money. My father used to say that we had the best canal in the country. It was never any trouble, it cost nothing to run and it always made a profit.' Thomas Hatchett, a lock-keeper in the canal's early days, might well have been envious; for thirteen years attendance at his lock he received no pay at all.

In its later years the canal was recalled by John Masefield, who lived at Ledbury as a child. He remembered 'the woman steering with her most becoming head-dress of a milkmaid's cap flapping on her cheeks, and the husband ahead minding the horse, and often singing, or knitting, or flinging words over his shoulder to the wife.' Sometimes he was asked to come aboard by 'these glorious people' and was shown 'the marvellous cabin, bright as a new pin, with its bunks and gear, the most wonderful abode on earth.' The canal ran alongside Bye Street which Masefield believed 'was their Sailor Town, and that there, when lamps were lit and little boys were in bed, they would make merry, and tell of the dangers of the deep.' When the railway was built on the line of the canal Masefield commemorated the event in his poem *The Widow in the Bye Street*. Here, six deep locks led up to the summit level – all demolished when the railway was built.

Masefield's recollections show that family narrow boats did use the Hereford & Gloucester, and a watercolour painting by Philip Ballard, brother of Stephen, shows a young lady with a large hat and a small dog on the stern of a narrow boat. On the cabin side panel is a decorative painting, and David Bick, the canal's historian, suggests that this is 'probably the earliest known illustration of a cabin with painted decoration'. But one hundred years after the Hereford & Gloucester Canal was closed a restoration society came into being, and in 1987 the first half-mile (0.8km) was formally reopened, leaving only another 33½ miles (54km) to go. And most of them are beautiful miles, too, the canal's course running through rich, rolling wooded country with obscure hidden tunnels, ruined lock sites and many of the canal's original bridges, including Stephen Ballard's splendid Skew Bridge at Monkhide, an engineering triumph tucked away in deepest Herefordshire.

The Leominster Canal, the Hereford & Gloucester's near neighbour, was even less successful – and even more obscure. It was intended to link the market towns of Kington and Leominster with the Severn opposite Stourport, thus connecting with the Staffordshire & Worcestershire Canal and the rest of the country. Coal from mines near its route at Mamble and Pensax was to be a principal cargo; but these mines, like those near Newent, were also soon exhausted. Only the central section, from Leominster to Marlbrook, was completed; Southnet Tunnel was left isolated and the great tunnel at Pensax and the flight of locks down to the Severn were never made. Funds could not be raised for the completion of its line. In 1858 it was bought by the Shrewsbury & Hereford Railway for £12,000, and soon afterwards it was closed and drained. The 18½ miles (29.4km) of its line were remote from habitation, and unless by chance you stumble upon a crumbling brick-built aqueduct across the River Rea, hidden in woodland, or odd dips and bumps in fields near Wooferton Station, it is hard to believe it was ever there.

Westward in Wales, however, are three country canals that are thriving: the Llangollen Canal in North Wales and the Brecknock & Abergavenny in the south are popular holiday waterways, while the Montgomeryshire Canal is being actively restored with several miles already reopened to boats. The first two were originally cut to serve the growing industry: the Llangollen was part of the Ellesmere Canal Company and supplied water as a navigable feeder; it also serviced the coal and iron fields near Ellesmere and Ruabon. The Brecknock & Abergavenny, with its companion canal, the Monmouthshire, was the central element in a network of horse-drawn tramroads bringing coal and iron ore from the hills to Newport, which soon developed into a major exporting port. With its massive aqueducts at Pontcysyllte and Chirk, the Llangollen is one of the most spectacular canal undertakings, while the Brecknock & Abergavenny, terraced on wooded hillsides above the River Usk, is 33 miles (53km) of almost uninterrupted beauty.

Crossing the Pontcysyllte Aqueduct represents the high point of hundreds of annual holiday voyages, but it is unlikely that today's navigators would experience the excitement and satisfaction of Tom Rolt when in the summer of 1949 he crossed the aqueduct in *Cressy*, the first narrow boat passage for at least ten years. He saw the aqueduct as a statement: 'At a time of its completion . . . the grandest embodiment of man's new confidence in his ability to master his environment, of man's faith in his own, rather than in any external, powers.' Within months of its opening in 1805 artists were making the difficult journey to sketch and paint the aqueduct in its dramatic setting, and more than a dozen engraved versions soon appeared to inform the public of this massive achievement. Chirk Aqueduct, too, had its illustrators, but some years later it was

The Monmouthshire Canal at Allt-yr-yn, near Newport, in the 1930s

overshadowed by a taller railway viaduct built alongside, epitomising the relative importance of these two means of communication in the mid-nineteenth century.

It is remarkable that the Llangollen Canal survived when other elements of what became the Shropshire Union Canal & Railway Company – the Newport branch, the Montgomeryshire Canal, the Shropshire tub-boat and the Shrewsbury canals – were abandoned. In 1944 the London, Midland & Scottish Railway, which had become the owners of the Shropshire Union system, sought the abandonment of 175 miles (281.5km) of canal, including the line to Llangollen – more accurately, to Llantisilio, 4 miles (6.4km) further on. However, because this served a purpose as a water supply channel – from the River Dee at Horseshoe Falls by Llantisilio – it was saved, and today even its short branches are being restored.

The survival of the Brecknock & Abergavenny Canal is equally remarkable. Alone of the South Wales valley canals – the Glamorganshire, the Swansea, the Neath, and its own companion the Monmouthshire Canal – it has remained fully open and navigable, even though the last toll was taken in 1933 and only two or three boats had been using the canal for the previous thirty years. Again, it survived because of its value as a water channel. Now, although isolated from the canal network, it has been renovated, and with five boatyards in 33 miles (53km) it is a busy little waterway in the summer.

In commercial days nearly all the traffic went through to Newport from the various wharves where the tramroads from the hills met the canal. At Pontymoile Basin, near Pontypool, boats moved on to the Monmouthshire Canal for the final 9 miles (14.5km) to the docks. In 1865 the Monmouthshire company bought the Brecknock & Abergavenny Canal for a little over £60,000, partly to safeguard the water supply. Then in 1880 the Great Western Railway took over both canals – the beginning of the end for the Monmouthshire which was closed bit by bit until the final length was abandoned in 1962.

There were eleven village wharves on the Brecknock & Abergavenny Canal, served until 1915 by a regular market boat from Newport, but otherwise the canal seems to have had little direct impact on the life of the countryside through which it ran. Limestone was burnt by some of the wharves, using coal brought by canal, and basic slag was brought from Govilon to Llangattock Wharf for distibution as fertiliser. In later years a boat would take village children on a Sunday School outing. Indirectly, however, because it cheapened the price of coal at Brecon and the villages nearby and enabled agricultural produce to be carried to market, the canal and its tramroad connections contributed to general development and a measure of prosperity over a wide area. It remains as a superb example of engineering skill with only six locks in this mountainous countryside, a magical waterway with a route that seems designed to explore the secret places, passing villages at rooftop height and with glimpses of wide open views across the Usk valley to the Brecon Beacons and beyond.

Of these three Welsh canals, the Montgomeryshire was the most purposefully agricultural, though at the time the canal was promoted it was said there was never any prospect of large profits accruing:

> . . . therefore the subscribers were the Noblemen and Gentlemen either possessed of estates in this County, or resident therein, who had for their object the Extension of Agriculture, the Reduction of Horses. . . the increase of Horned Cattle, and the Preservation of the Roads; with the consequent Advantage to the Public.

The Montgomeryshire was cut from a junction with the Llangollen Canal at Frankton, through Welshpool, to its terminus at Newtown;

its chief purpose was to carry lime from quarries at Llanymynech
to the settlements and farms along its route, and at one time there
were ninety-two kilns alongside the canal. Coal, timber, grain, dairy
produce and general merchandise were also carried, and in the
mid-nineteenth century there were thirty warehouses in its 35 miles
(56km) length. In *The Archaeology of the Monmouthshire Canal*
published by the Royal Commission on Ancient and Historical
Monuments in Wales (1981) Stephen Hughes describes the canal as
the artery of the landscape:

> . . . providing transport and sometimes power to an inter-
> dependent series of industrial and agricultural installations.
> Branching out from the waterway into the surrounding hinter-
> land were the routes of the carriers and waggoners busying
> themselves with delivering lime for the fields and collecting the
> harvest of fine Montgomeryshire oak.

Four important settlements developed alongside the Montgomeryshire Canal in its early years. Two were extensions of existing
towns: Newtown, where there was a strong woollen industry and
where the canal company built a large dock with wharf buildings and
warehouses, workers' cottages and limekilns; and Welshpool, which
became the headquarters of the canal's administration. Then there
was Pool Quay, which antedated the canal by more than a hundred
years as it marked the head of navigation on the River Severn. Here
the 'fine Montgomeryshire oak' was loaded on its way to the naval
dockyards. The canal brought added trade to Pool Quay, with goods
being trans-shipped from canal to river for conveyance to the main
inland waterway network through Stourport; this diminished, however, when the canal itself was linked to the main system. Looking
at Pool Quay on the Severn today it seems extraordinary that sailing
vessels ever traded from there; of the once-busy wharves only part of
a single building remains. No towing-path was ever built along the
Upper Severn and boats had to be bow-hauled when the wind was
contrary or very light.

The fourth settlement was Garthmyl, for some twenty years
the terminus of the canal until money became available for its
continuation to Newtown. Here a small canal port developed, traces
of which may still be seen – there were two wharf houses, two warehouses, several wharves including a coal wharf, an office, stables and
banks of limekilns. The Nag's Head, successor to a much older pub,

was built in the canal's early days and a brewhouse and maltings were added later. As Newtown developed Garthmyl's importance declined, but it is still one of the very few surviving examples of a rural canal port.

Daily life on the Montgomeryshire Canal had its excitements, especially at the locks. On 12 January 1827 John Bagley hurled a brick at the Carreghofa lock-keeper, Edward Perkin, injuring him so severely that he died a week later. Why? – perhaps it was an argument about damage to the lock gates, and maybe there was some provocation as eighteen-year-old Bagley was found guilty not of murder but of manslaughter and was sent to prison. Carreghofa locks were not lucky for their keepers; a few years later Perkin's successor lost his job and cottage when he was found guilty of wrongly gauging a boat. Several boatmen were fined for damaging the locks and drawbridges and Mr Hill, the agent, declared that 'unless there are some able stout men to look after and take care of the locks they will soon all be destroyed.' The canal company was compelled to build cottages for their lock-keepers – something they had not originally intended to do – and to pay them extra (just 2s 6d (12½p) a week) for keeping the locks open at night. The lock-keepers sometimes had to deal with difficult customers, such as Mr Salmon, a boatman employed by the British Iron Company, 'who got himself so much intoxicated as to be unable to steer his boat'; and Peter Lewis, a boatman working out of Newport, who left a dead horse on the towing-path.

Daily life is very different on the Montgomeryshire Canal today. The stretch between Welshpool and Burgeddin is navigable through four locks, restored to use with funds raised by the Prince of Wales' Committee; here the boat *Heulwen-Sunshine* may be seen, bought by the ladies of the Inland Waterways Association for the use of disabled people. Walkers following the Offa's Dyke Path might be using the towing-path near Buttington, because here the two coincide; and at Frankton and Carreghofa teams of volunteers are a familiar sight, repairing and restoring the locks. Rednal Basin was once an interchange point for passengers taking the canal from Newport and the train from Rednal Station on the Shrewsbury–Chester line; here, botanists might be working in the nature reserve, and similarly at the winding hole at Abbey Bridges, or along the short Guilsfield Arm. With roach, bream, tench, carp and pike to be caught, anglers are frequent visitors to the canal banks. Industrial archaeologists may be following in the footsteps of Stephen Hughes. And as there are several Sites of Special Scientific Interest along its length, the canal attracts naturalists, ornithologists, specialists of all kinds – besides all those who simply love the open air and the variety of wildlife that the canal supports.

Full restoration of the Montgomeryshire Canal is questioned, even opposed, by the many individuals and organisations whose interest lies in nature conservation, on the grounds that the necessary dredging and the passage of boats will destroy what they value most. But this is already a managed environment; the canal has a conservation officer based at Llanymynech, and many of the special sites have been established off the main channel. Once the canal is reopened changes are bound to occur; but changes will occur anyway through the effect of natural forces.

However, conservation and navigation can and do co-exist; for example the Crinan Canal in Scotland is an 11-mile (17.6km) waterway that cuts through the northern end of the Kintyre peninsula, with about 3,500 craft passing through every year: sturdy fishing boats, elegant and expensive yachts, powerful motor cruisers made for the open sea – only narrow boats are missing. Yet about half the length of the Crinan Canal is a designated conservation area; there

An otter inspecting the Crinan Canal (*Waterways News*)

is a wide variety of bird life, including birds of prey – golden eagles, kestrels, sparrow hawks and buzzards – and otters, which are seen swimming in the canal perhaps looking for sea and brown trout. British Waterways staff have provided twenty-five nesting boxes alongside, and avoid cutting the grass on the banks where butterfly orchids flower. There is also a rich insect population with several rare butterflies. Furthermore the countryside of the Crinan Canal is superb: to the south, Knapdale Forest, to the north the mysterious expanse of Moine Mhor, dominated by the rocky outcrop of Dunadd, once the fortified headquarters of the ancient kingdom of Dalriada. And there can be few more beautiful places on any canal than Bellanoch, where the Crinan Canal widens into a bay or lagoon and the views stretch out across the estuary of the River Add to the islands in the Sound of Jura.

In recent years it has been claimed that the interests of nature conservationists and inland navigators are irreconcilable. Often these differences are voiced quite violently, as when the reopening of upper reaches of the Yorkshire Derwent was opposed. This need not be so – the Crinan and the Montgomeryshire canals prove the opposite. Where the authorities involved are willing to work together – British Waterways Board, the Nature Conservancy Council, the various organisations and societies whose members wish to use and enjoy the waterway – the different interests can be reconciled and happy solutions can be found. It is good to end this study of the country canal on such a hopeful note.

Napton bottom lock and lock cottage on the Oxford Canal (British Waterways Board)

Bibliography

Allsop, N. *The Somersetshire Coal Canal Rediscovered* (Millstream Books, 1988)

Anon (H. Schofield). *The Waterways to London* (Simpkin Marshall, 1869)

Astbury, A. K. *The Black Fens* (SR Publishers, 1970)

Bick, D. *The Herefordshire & Gloucestershire Canal* (Pound House, 1979)

Bloom, A. *The Farm in the Fen* (Faber, 1944)

Clew, K. *The Somersetshire Coal Canal & Railways* (David & Charles, 1970)

Clew, K. *The Dorset & Somerset Canal* (David & Charles, 1971)

Cove-Smith, C. *The Grantham Canal Today* (Grantham Canal Society, 1986)

Dalby, L. J. *The Wilts & Berks Canal* (Oakwood Press, 1986)

Darby, H. C. *The Draining of the Fens* (CUP, 1968)

Denton, J. H. *The Montgomeryshire Canal* (Goose & Son, 1984)

Ewans, M. C. *The Haytor Granite Tramway and the Stover Canal* (David & Charles, 1964)

Harris, H. *The Grand Western Canal* (David & Charles, 1973)

Harris, H. & Ellis, M. *The Bude Canal* (David & Charles, 1972)

Hassell, J. *A Tour of the Grand Junction Canal* (J. Hassell, 1819)

Hoskins, W. G. *English Landscapes* (BBC, 1973)

Household, H. *The Thames & Severn Canal* (David & Charles, 1969)

Hughes, S. *The Archaeology of the Montgomeryshire Canal* (RCAHM Wales, 1983)

Jackson, P. *Waterways & Wetlands* (Philip Jackson, 1987)

Messenger, M. J. *Caradon & Looe* (Twelveheads Press, 1978)

Mingay, G. E. *Rural Life in Victorian England* (Heinemann, 1977)

Rendell, J. *Along the Bude Canal* (Bossinney Books, 1979)

Rendell, J. *Story of the Bude Canal* (Stannary Press, 1987)

Rolt, L. T. C. *Narrow Boat* (Eyre Methuen, 1944)

Russell, R. *Lost Canals & Waterways of Britain* (David & Charles, 1982)

Smith, G. *Our Canal Population* (Haughton & Co, 1878)

Smith, G. *Canal Adventures by Moonlight* (Hodder & Stoughton, 1881)

Steven, R. A. *Brecknock & Abergavenny and Monmouthshire Canals* (Goose & Son, 1974)

Thurston, E. Temple *The Flower of Gloucester* (William & Norgate, 1911)

Vine, P. A. L. *London's Lost Route to the Sea* (David & Charles, 1965)

Warner, P. *Lock Keeper's Daughter* (Shepperton Swan, 1986)

How traditional can you be? (see page 83)

Acknowledgements

I am grateful to those who helped in different ways with *The Country Canal*, especially to Philip Jackson for his evocative painting reproduced on the dust jacket, Ulla Frisch for her crisp drawings and Yvonne Barnett for her cheerful cartoons. Other help was given by David Bick, Charles Hadfield, the Archivist of the Waterways Museum at Stoke Bruerne, Croydon Central Library, Huntingdon Record Office, County Record Office Truro, G. M. Hollingston and the West Country Branch of the Inland Waterways Association, the Curator of Newbury Museum, Sheila Doeg and *Waterways News*, British Waterways Board Photographic Collection, the Cambridgeshire Collection, and Simon Ross. For permission to use quotations from *Lock Keeper's Daughter* my thanks are due to Pat Warner and Shepperton Swan, and from *The Archaeology of the Montgomeryshire Canal* to Stephen Hughes and the Royal Commission on Ancient & Historical Monuments in Wales. The photographs, except where otherwise acknowledged, and all opinions and errors are my own.

Index

Adventurer's Fen, 101, 103
Andover Canal, 139
Anglesey Abbey, 107
Aqueducts: Almond, 30; Avon, 30, 32; Barton, 34; Beam, 124; Brynich, 28; Chirk, 32, 148; *31*; Dundas, 28–30; *29*; Dunkerton, 122; Lune, 30; Marple, 30; Midford, 122; Murtry, 123; Pontcysyllte, 148; *32*; Slateford, 30
Astbury, A. K., 88–9
Avon Industrial Building Trust, 122

Ballard, Stephen, 145, 147
'Bargee Ware', 51
Benwick, 86–91
Bethell, Mrs Charlotte, 143
Bick, David, 147
Birket Foster, Miles, 7, 45
Bloom, Alan, 103–4
Blue Line Canal Carriers, 55
Boatman's Friend Society, 54
Bottisham Lode, 107
Brecknock & Abergavenny Canal, 10, 14, 28, 40, 129, 149–50; *9*, *39*
Bridgwater & Taunton Canal, 126–7; *126*
Brindley, James, 34, 58, 63
Broke, Rev H. G., 92–4
Bude Canal, 11, 12, 26, 110, 111–13; *109*, *118*
Burwell, 102; Lode, 101–104
Bury Brook, 91–2

Caistor Canal, 143
Caldon Canal, 8–9, 14; *29*
Caledonian Canal, 21, 23–4, 32; *31*
Cam, River, 98–9
Cambridgeshire Navigable Lodes, 99–107
Canal Boats Acts, 54
Canal construction, 62–3; *17*
Casaubon, Isaac, 90
Census returns, 53
Chard Canal, 113–14
Commercial End, 106–7
Conservation, 153–4
Constable, John, 45; *44*
Coprolites, 105

Crinan Canal, 153–4; *153*
Crosley, William, 40
Croydon Canal, 43; *10*, *26*

Darby, Prof H. C., 37, 96
Denver Sluice, 97
Devil's Dyke, 104–5
Dorset & Somerset Canal, 109, 115, 123–4
Drainage, 36–7
Driffield Canal, 14, 144; *142*

Edyvean, John, 119
Ely, 98, 106
Exeter Ship Canal, 126

Fenland Waterways, 34–8, 86ff; *33*
Fielder, Richard R., 99–100
Floating church, 94
Fussell, James, 115

Garthmyl, 151
Glastonbury Canal, 123
Gloucester & Berkeley Canal, 21
Grand Junction/Grand Union Canal, 15, 17–19, 21; *17*, *41*
    Buckingham Branch, 17–19
    Wendover Arm, 26–7
Grand Western Canal, 111, 115, 116–17; *112*, *116*
Granite trade, 18
Grantham Canal, 10, 130–1
Great Ouse, River, 97
Green, James, 109–10, 114, 115

Hanging plates, 52
Hassell, John, 42
Hay inclined plane, 114
Hereford & Gloucester Canal, 12, 20, 21, 82, 145–8; *13*, *146*
Hobbacott Down inclined plane, 110, 111
Hollingshead, John, 21, 42
Holme, 92
Holme Fen posts, 88
Hoskins, Prof W. G., 38
Hughes, Stephen, 151

Ice boats, 24–5

Illustrated London News, 45–6; *48*, *50*, *52*, *54*
Ireland: Grand Canal, 23; Royal Canal, 23; Barrow Navigation, 23
Ironbridge Open Air Museum, 114

Jackson, Vic, 106; *105*
James, Jack, 55

Kennet & Avon Canal, 10, 12, 14, 20, 28, 135–7; *13*, *29*, *44*, *136*

Lancaster Canal, 30
Leader, Benjamin, 45
Leicester Navigation, 18, 38, 49–50; *19*
Lengthsmen, 24
Leominster Canal, 10, 14, 129, 147–8
Leven Canal, 143
Liskeard & Looe Canal, 120
Llangollen Canal, 34, 148–9; *31*, *32*
Lock cottages, 21
Lock keepers, 20–4, 152
Louth Canal, 141; *140*

Macclesfield Canal, 34, 40
March, Cambs, 94–5
Market Weighton Canal, 144–5
Masefield, John, 147
Measham ware, 51
Middle Level, 86ff
Mole-catcher, 25; *25*
Monmouthshire Canal, 28, 150; *149*
Montgomeryshire Canal, 10, 14, 150–3
Morwellham Open Air Museum, 110, 125
Mountsorrel, 18–19; *19*

Navigators, 62
Nene Navigation, 26
Nene, Old River, 86ff
Nix, Mrs Sheila, 143, 145

Orford, Lord, 90, 97
Outwell, 95–6; *95*
Oxford Canal, 21; *130*, *154*

Pocklington Canal, 143–4; *142*
Pool Quay, 151

Railway mania, 43
Ramsey, 91–2; *93*; Mere, 91; Forty Foot, 91
Reach, Cambs, 104–6; Lode, 101–4
Rennie, John, 28–30
Restoration, canal, 131, 152
Rolt, L. T. C., 41, 76
Royal Military Canal, 139–40

Sapperton Tunnel, 134–5; *133*

Schofield, Harold, 44, 135
Shropshire Union Canal, 21
Siddons family, 26
Sleaford Navigation, 16
Smale family, 26
Smith, George, 47ff
Somersetshire Coal Canal, 20, 108, 114, 121–3; *120*
Southey, Robert, 47
Southnet Tunnel, 134
Staffordshire & Worcestershire Canal, 28
Stamford Canal, 141
St Columb Canal, 117, 119
Stockwell, Gen. Sir Hugh, 135
Stover Canal, 125
Stretham Old Engine, *33*
Swaffham Bulbeck Lode, 106

Tamar Manure Navigation, 119, 126
Tavistock Canal, 125–6
Telford, Thomas, 30–4
Thames & Medway Canal, 139
Thames & Severn Canal, 12, 20, 21, 44, 131–5; *133*
Thurston, E. Temple, 41–2
Torrington Canal, 117, 124–5
Tramroads, 14
Trent & Mersey Canal, 28
Tub-boat canals, 134–5
Tunnels, 134–5

Upware, 99–100; Republic, 99–100
Upwell, 95–6; *93*

Vermuyden, Cornelius, 34, 36–7
Vine, Paul, 138, 139

Warner, Pat, 21–2
Watering meadows, 60
Waterways Museum, 55
Watts, Frederick W., 45
Weldon, Robert, 114
Well Creek, 96–7
Wellstream, 96
Westport Canal, 126
West Water, 89
Wey & Arun Junction Canal, 20, 138–9
Wharves, 11ff; *13*
Whittlesey Mere, 92
Wicken Fen, 38, 101; Lode, 100–1, *100*
Willmer, Prof E. N., 101
Wilts & Berks Canal, 11, 12, 138
Wisbech Canal, 96
Wisbech & Upwell Tramway, 96